PURCHASING
FOR
FOODSERVICE

SECOND EDITION

PURCHASING

OR FOODSERVICE

SECOND EDITION

Lynne Nannen Robertson

PhD, REGISTERED DIETITIAN

Iowa State University Press / Ames

Lynne Nannen Robertson, PhD, RD/LD, was assistant director, Department of Dietetics, Iowa Methodist Medical Center; director, Food Services and Food Service Programs, Des Moines Area Community College; and coordinator of instruction, Department of Nutrition and Dietetics, University of Missouri-Columbia Medical Center. She is now president of Creative Concepts Foodservice Consultants.

©1994 Iowa State University Press, Ames, Iowa 50014
All rights reserved

Authorization to photocopy items for internal or personal use, or the internal or personal use of specific clients, is granted by Iowa State University Press, provided that the base fee of $.10 per copy is paid directly to the Copyright Clearance Center, 27 Congress Street, Salem, MA 01970. For those organizations that have been granted a photocopy license by CCC, a separate system of payments has been arranged. The fee code for users of the Transactional Reporting Service is 0-8138-1463-4/94 $.10.

♾ Printed on acid-free paper in the United States of America

Originally published in 1985 as *Purchasing for Food Service: Self-Instruction* by Lynne N. Ross through three printings (1985–1993)

This edition, 1994

Library of Congress Cataloging-in-Publication Data

Robertson, Lynne Nannen
    Purchasing for foodservice / Lynne Nannen Robertson.—2nd ed.
        p.   cm.
    Prev. ed.: Purchasing for food service. 1st ed. 1985.
    Includes index.
    ISBN 0-8138-1463-4 (alk. paper)
    1. Purchasing   2. Foodservice.   I. Robertson, Lynne Nannen. Purchasing for foodservice.
II. Title.
    TX911.3.P8R67 1994
    647.95′0687—dc20   94-5798

CONTENTS

PREFACE *vii*

1. THE PHILOSOPHY OF PURCHASING *3*

2. ACCOUNTING PROCEDURES *10*

3. BEEF *23*

4. PORK *34*

5. LAMB *40*

6. SEAFOOD *43*

7. POULTRY *51*

8. EGGS *57*

9. DAIRY PRODUCTS AND ALTERNATIVES *63*

10. CHEESE *69*

11. FRESH FRUITS *75*

12. FRESH VEGETABLES *87*

13. CANNED, FROZEN, AND DRIED FRUITS AND VEGETABLES *104*

14. CEREAL PRODUCTS *108*

15. SPICES, HERBS, AND FLAVORINGS *120*

v

16. SWEETENING AGENTS 128

17. BEVERAGES 135

18. RECEIVING AND STORAGE 143

INDEX 147

PREFACE

IN THIS TIME OF HIGH COSTS, administrators are looking at every possible way to economize without adversely affecting foodservice quality and the welfare of those served. Purchasing techniques can be very useful in maximizing the resources of the dietary department. This book will assist you in learning accounting procedures and product specifications that will be needed by the competent purchaser in performing the duties necessary to provide optimum purchasing decisions basic to a successful foodservice operation. The foods included are those used in most foodservice departments. No attempt has been made to cover all foods available.

This book may be used by any level of foodservice personnel that is involved with a phase of the purchasing process. It is written to be understood by all. It would be desirable for the book to be used with the guidance and direction of a dietitian or instructor, but any interested employee working alone should benefit from the information.

Purchasing for Foodservice has been organized to simplify a rather complex subject. It is hoped that this information will serve as a tool for making improvements in the purchasing function of foodservice departments.

PURCHASING FOR FOODSERVICE

SECOND EDITION

1 THE PHILOSOPHY OF PURCHASING

THE PURCHASE OF FOOD AND SUPPLIES for a foodservice department is a large responsibility. To get the best value for the money spent takes education, care, and planning. Parameters that need to be considered in making basic purchasing decisions are the type of institution, production needs, shelf life of food, general market and economic conditions, amount of storage (dry, refrigerated, and frozen), closeness to supply, and availability of capital.

ORDERING

The menu represents the type of food that the administration wants served in the facility. Therefore, an ordering system must be put into effect to assure that the right quantity of food is on hand to correspond with that menu. The foodservice may be a country club, a school, a college or university, an industrial cafeteria, a commercial restaurant, or a health care facility.

The first step necessary to set up an ordering system is a complete menu, which includes portion sizes and detailed food item descriptions: for example, *sugar cookie* instead of merely *cookie* or *pineapple upside-down cake* instead of *cake* (see Ch. 2, Forms B and C). The menu may be written for each individual day and used only once, or it may be written in cycles of three, four, or five weeks depending on the policy of the facility.

After assembling the menu the next task is to check the recipes for ingredients and to prepare a list of required items. Another list should be prepared for staples such as flour, sugar, spices, paper products, and cleaning supplies. Determine the amount of food and supplies needed to last from one delivery date until the next. This list should be checked against the inventory to determine the items and quantities needed.

Foods should be selected according to the intended use. If a higher quality is purchased than is required, unnecessary additional cost will result. To assist in avoiding this error, careful specifications should be prepared. Specifications are descriptions of items and should be in as much detail as possible. They include, but are not limited to, the name of the product, size of the container or unit, condition of the product (whole, sliced, cut, diced, peeled), and number of pieces, grade, size, shape, weight, and source of the product.

The federal Food, Drug, and Cosmetic Act requires that labels for any food passing from one state to another include the following information:

1. The name of the product and the variety or style, when applicable.

2. The name and address of the manufacturer, packer, or shipper.

3. An accurate statement of the quantity.

4. A list of the common or usual names of each ingredient when two or more foods go into a product that is not standardized. The order of listing indicates the relative amount of each food; the food present in the largest quantity comes first.

5. Enough facts to fully inform the purchaser concerning foods for special dietary purposes.

6. A statement of the use of artificial flavoring, artificial coloring, imitations, or chemical preservatives.

7. An indication and explanation of substandard quality or substandard fill of container. Such food must be wholesome, even though it does not measure up to prescribed minimums in a stated way.

8. The required information in an easy-to-find and easy-to-understand statement.

Some states have established their own regulations, which are usually higher than federal standards. For example, Washington state has a grading system for apples that is stricter than the federal grading system.

FOOD SAFETY

Food safety is a major concern for food buyers; therefore, it is necessary to assure that the food purchased is safe and that it will be held, prepared, and served in a safe manner. The only way to eliminate bacteria from meat is to use irradiation, or ionizing energy. Irradiation exposes food to a dose of gamma rays from radioactive cobalt. Depending on the dose, it can kill bacteria, insects, and fungi; it can also retard ripening and spoilage. Irradiation of foods is the proven safe and effective means of breaking the cycle of food-borne disease. It is currently used for this purpose in more than 30 countries around the world. Once meat is irradiated, not only are life-threatening bacteria destroyed, but meat can be eaten rare or stored overnight without refrigeration if necessary. Irradiation does not make objects radioactive. Although antitechnology advocates are circulating unfounded health hazard claims, food irradiation has been endorsed by the World Health Organization, United States Department of Agriculture, National Food Processors Association, American Council on Science and Health, and Food and Drug Administration.

COST COMPARISONS

Price comparisons are important, but the acceptability of a product should be considered as well when decisions are made about which brand to purchase. Quantities and prices of food products follow seasonal variations. They are best in quality and lowest in price at the peak of the growing season. Price may also be influenced by other factors such as promotional discounts, large quantities, coop purchasing, discontinued items, night or off-hour deliveries, introductory offers for new products, discounts for cash payment, or lower-quality products.

It is advisable to compare the products and prices of all acceptable vendors, being sure all products are of the same specifications. When holding a can cutting to taste and compare products, it is recommended that the government standards given earlier be used as guidelines. These take into account characteristics such as color, texture, flavor, maturity, absence of defects, and drained weight.

For cost comparison, entrees must include equal amounts of

protein or the ingredients necessary to make them equal. The cost of convenience foods should be compared to the cost of ingredients and labor for preparing conventional items. Rising labor costs have done much to promote the use of convenience foods—dehydrated, freeze-dried, canned, and frozen. The purchasing decisions related to labor need to include factors such as cost of labor required for the two modes of preparation, availability of labor at the skill level required, and availability of enough employees to prepare the menu items.

Frequency of delivery may also be an influencing factor. The delivery schedule needed and/or offered is affected by the size of the operation, the distance to be traveled, the products being purchased, and the availability of storage space—both room-temperature storage for dehydrated and freeze-dried products and low-temperature storage for refrigerated and frozen products.

Appearance, another factor in acceptability, is indicated by uniformity of pieces in cans of fruits, vegetables, and soups; consistency in gravies and sauces; and color in fruits and vegetables.

Acceptability of a new item should be checked against that for other brands, recipes, or forms of the food. Conduct a taste test, or check the amount of plate waste on the day the new item is on the menu, and compare it to the day the regular item is served. When testing a new item, purchase only a small amount until cost and acceptability have been checked.

Purchasing methods vary with the size and type of operation. The larger and more complex the operation, the more formal and complex the purchasing procedures.

FORMAL BUYING

For large facilities formal buying is usually done in writing. The potential suppliers are given the specifications and invited to bid. Can cuttings and product comparisons examine characteristics such as count, size, syrup density, color, flavor, and texture. The samples are checked and compared by several individuals. The information is recorded and tabulated on a score sheet (see Form A at the end of this chapter) before a final decision is made, keeping in mind usage and cost per serving. Can cuttings and product comparisons are held before the contract or agreement is finalized. Inventory purchases and items to be stored, such as canned fruits and vegetables and staples, are the things most likely to be purchased formally. These items are placed in the storeroom or cold storage areas when they are checked in. The decisions about how much to buy and how often are based on the Economic Order Quantity formula. However, the small operator may find it impractical to establish specifications and to use such an elaborate system. This decision is usually based on availability of capital, amount of storage space, and volume of business. The smaller operators may try several brands of a food item and then select the one they prefer.

OPEN OR INFORMAL BUYING

Open or informal buying negotiations are usually oral. The contract may be made either in person or by telephone. The person responsible for purchasing telephones suppliers who have copies of the operation's specifications to get prices on the items needed on a particular day. Price comparisons can also be made by using the price printouts received from each potential supplier. After the purchaser gets the prices and selects the purveyors, the order is placed specifying quantities and delivery dates. Usually the order is awarded to the supplier that has the lowest bid on a particular item. However, exceptions may occur when one supplier has the low bid on only a small fraction of the order or when several items are only a few cents lower than the price quoted by the supplier with the major portion of the order. If this occurs, it may be better to get the entire order from a single supplier rather than split the order. Although price is the major concern, service is also an important consideration. If specifications are not met, or deliveries are not made on time, or emergency service is refused, the purchaser may be justified in not continuing to order from a purveyor.

Food items such as dairy products, bread and buns, or coffee may be supplied by the vendor, based on a previous agreement (called "par stock") to keep up an established inventory. A standing order may be set up for items that are generally used at a consistent rate, and an established amount will be delivered on a predetermined schedule. If the inventory gets too high or too low, spot adjustments are made on a one-time basis to bring the on-hand amount back into line.

Management should be certain that the qualities and quantities invoiced have been received.

Direct purchases bypass the storeroom and go directly into production from receiving. Purchase these items as close to the time of use as possible to assure freshness. Items purchased this way include fresh fish and produce that are purchased for a particular menu listing.

PURCHASING CONTROLS

Purchasing controls consist of purchase orders or purchase requisitions, spot checks of the quotation sheets used in informal purchasing to be sure the items with the lowest prices are being selected, dollar distribution analysis to be sure all suppliers are receiving a fair share of the business, spot checks of inventory to assure par stocks are adequate, and comparison of prices with other competitors to assure the best prices are being quoted.

FORM A

Factors	Max. Points	Fancy	Ex. Std.	Standard
	ITEM			
	(Cut Green or Wax Beans)			
Clearness of Liquor	10	9–10	8	7
Color	15	14–15	12–13	10–11
Absence of Defects	35	31–35	28–30	25–27
Character*	40	36–40	32–35	28–31
Minimum Score		90	80	70

Factors	Can #1	Can #2	Can #3	Can #4	Can #5	Can #6
Clearness of Liquor						
Color						
Absence of Defects						
Character						
Total Score						
Drained Weight						
Liquid						

COMMENTS:

*For an item such as cut green or wax beans: good character (Grade A) indicates beans very young, tender (no more than 5% with tough strings), and full fleshed, with seeds in early stages of maturity.

2 ACCOUNTING PROCEDURES

EFFICIENT FOODSERVICE MANAGEMENT is based on effective accounting procedures. Any accounting experience with food purchased in the past is a valuable resource to use in making decisions for future buying.

BUDGET

A budget is a projected plan for expenditures, based on a record of past expenses (see Form B). It is used as a measuring stick to evaluate current performance and as a guideline for future spending. The proposal for expenditures may be for whatever time period the facility chooses and may be grouped into the categories that seem most logical to those using the information.

MENU

A menu is a plan or record of food items to be prepared and served by the personnel of the foodservice department (see Forms C and D). The items may range from a very simple, single-selection menu offered once each day to a very elaborate multiple-offering menu in a facility that is open 18 to 24 hours each day. The format used in each facility needs to be developed for and by that facility.

FOOD INVENTORY

A food inventory is an organized listing of all food on hand. It is used to determine amounts of food to order and as a check on the amount of food that should be stored. The inventory sheet may list specifications for each food item (see Form E). Items on an inventory should be grouped according to category and be in alphabetical order within each. It is usually convenient to have the storeroom arranged in the same manner. Shelves that are labeled will help assure that items are always placed in the same location. If the cost per purchase unit and the cost per issue unit are kept on the inventory form, time spent in determining the value of the inventory can be greatly reduced.

PRICE QUOTATIONS

A price quotation sheet is used to compile price information about food items from several vendors (see Form F). It is often used for the purchase of meat items for which the price varies from week to week, and the savings realized by a few cents variation may make a large difference in expense. After the item and amount have been determined, quotations are requested from vendors. Each vendor's price is recorded in the appropriate column. Based on the quotations, a vendor is selected for each item needed.

PURCHASE ORDERS AND RECORDS

Purchase orders should be used to assure the receipt of items as specified. A purchase order provides a written record of the exact amounts and specifications of the food items ordered and the prices that are expected on the invoice. Such a form provides a record for the salesperson receiving the order, the buyer placing the order, the receiving clerk checking in the order, and the accounting office person paying the invoice.

A purchase record is a summary of the cost of all deliveries received in a specified period (see Forms G and H). In a large facility it may be totaled and transmitted daily, while in a small facility it may be kept for an entire month before summarizing. Some managers prefer to divide receipts by vendor; others choose to categorize by food classification.

MEAL CENSUS

A monthly meal census is a record of all meals served each day and is divided into appropriate categories according to the facility (see Form I). In a private club the categories may be members, guests, and employees. In a school foodservice the categories may be students, faculty, and employees. In a health care facility the divisions might be patients, employees, guests, and free meals. The record of meals served is necessary to calculate food, labor, and total cost per meal.

FOOD, MEAL, AND PORTION COSTS

Food costs and meal costs are calculated using the following formulas: $BI + P - EI = C$, where BI = beginning inventory, P = purchases, EI = ending inventory, and C = food cost. In a commercial operation food cost percentage is found by $C \div S = C\%$, where C = food cost, S = sales, and $C\%$ = food cost percentage. In student feeding and health care facilities the cost per meal is found by $C \div M = MC$, where C = food cost, M = meals served, and MC = cost per meal. If adjustments are made for employee meal allowances, sales of food for income, or free meals for public relations, these amounts need to be subtracted before calculations are made.

The portion cost is found by dividing the total cost by the number of servings (see Form J). The cost of a serving of fruit or vegetables is found by dividing the cost of the can by the number of portions it provides. The cost of a recipe item is found by adding the cost of each ingredient and dividing the total by the number of portions it yields.

MONTHLY SUMMARIES

The monthly summary is a compilation of all the information obtained through preparing and summarizing all other foodservice forms (see Form K). Each summary provides useful information about the operation for that month when compared to the budget, and a collection of several summaries can show any trends that are developing in the operation. This may be the quickest and surest way of spotting management problems so that corrective action can be taken before the problem becomes too acute.

FORM B

FOOD SERVICE DEPARTMENT BUDGET

19_____

Item	Annual	Monthly
Salaries and wages	_____	_____
Food	_____	_____
Disposables	_____	_____
Cleaning supplies	_____	_____
Other expense	_____	_____
Education	_____	_____
TOTAL	_____	_____

FORM C

MENU

Sunday	Monday	Tuesday	Wednesday	Thursday	Friday	Saturday
Oatmeal	Cream of Wheat	Malt-o-Meal	Oatmeal	Cream of Wheat	Oatmeal	Malt-o-Meal
Scrambled egg	Fried egg	Soft-cooked egg	Scrambled egg	Poached egg	Soft-cooked egg	Scrambled egg
Sweet roll	Toast	Toast	Toast	Toast	Toast	Pancakes
Assorted juices	Assorted juices	Prunes	Grapefruit	Assorted juices	Assorted juices	Assorted juices
Roast beef	Chicken loaf	Pork chop	Tater-Tot casserole	Swiss steak	Meat loaf	Baked fish
Whipped potato	Cream-style corn	Parslied potato		Baked potato	Mashed potato	Hash browns
Peas and carrots	Green beans	Broccoli	Harvard beets	Buttered carrots	Mixed vegetables	Stewed tomatoes
Ginger ale salad	Relish tray	Aspic salad	Cranapple juice	Head lettuce	Relish plate	Coleslaw
Cheesecake	Peaches	Ice cream	Baked custard	Fruit cup	Purple plums	Brownie
	Ham-rice casserole	Cream of tomato soup	Escalloped chicken	Zucchini casserole	Corn chowder	Hot turkey sandwich
	Spinach	Assorted crackers	Buttered peas	Cottage cheese	Assorted crackers	French fries
	Pineapple and cottage cheese salad	Egg salad sandwich	Sunshine salad	Applesauce	Cheese plate	Squash
	Hot rolls	24-hour salad	Hot rolls	Ice cream	Waldorf salad	Cranberry jelly
	Frosted creams	Gingerbread	Chocolate pie	Sugar cookies	Caramel pudding	Pears

FORM D

COMPUTER ASSISTED MENU PLAN

REGULAR DIET
```
           8/20    WEDNESDAY
         COST  BREAKFAST      (CATEGORIES  1- 3)
    2042  6.6  TOMATO JUICE
    3984 14.1  SCRAMBLED EGGS
    1455 10.7  SHREDDED WHEAT

         COST  DINNER        (CATEGORIES  4- 8)
    3396 15.5  MACARONI AND CHEESE
    5829 10.2  PEAS, FROZEN, BU
    6242  8.4  COLE SLAW
    7803 15.3  BREAD PUDDING

         COST  SUPPER        (CATEGORIES  9-14)
    3863 17.6  SAUTEED CHICKEN LIVERS
    4205  9.1  BAKED POTATO
    5322  8.9  GREEN BEANS, CND, BU
    6122 13.5  LETTUCE WEDGE / 1000 ISLAND
    7256 12.7  CHOCOLATE CHIP COOKIE (ONE)
```

TOTAL COST= $1.426; COST INCLUDING BREAD & BEVERAGE= $1.828
TYPE GO SET STOP LIST ENTER FORCE MENU NUTRIENTS OR RESTART
NUTRIENTS

NUTRIENT COMPOSITION:

	CAL.	PROT. GR.	FAT GR.	CHO. GR.	FIBER GR.	ASH GR.	CALC. MG.	PHOS. MG.	IRON MG.	SOD. MG.	POTAS. MG.	VIT A I.U.	THIA. MG.	RIBO. MG.	NIAC. MG.
	1631	66.5	61.9	194.	5.59	17.	949.	1360.	13.1	2617.	2394.	6491.	0.97	2.	12.3
W/B+B	2318.	89.4	94.2	270.	5.73	22.	1561.	1869.	15.3	3395.	3201	7696.	1.28	3.	15.1

TYPE GO SET STOP LIST ENTER FORCE MENU NUTRIENTS OR RESTART
GO
AVERAGE COST FOR 1. DAYS= $1.828 (INCLUDING BREAD & BEVERAGE)

FORM E

INVENTORY

Canned vegetables	Size of case	Cost per case	Cost per can	Size of serving	Servings per can	Cost per serving	Amount on hand	Value on hand
Asparagus spears (55–78)	6 #5			3 or 4	20			
Beans, cut green, Blue Lake, 3 sieve	6 #10			3 oz	33			
Beans, cut wax, 3 sieve	6 #10			3 oz	30			
Beans, kidney	6 #10			3 oz	35			
Beans, baked	12 #5			4 oz	27			
Beet, whole pickled	6 #10			3 oz	33			
Beets, sliced	6 #10			3 oz	33			
Beets, sliced pickled	6 #10			2 oz	50			
Beets, tiny whole (80–90)	6 #10			4 pc	25			
Carrots, sliced	6 #10			3 oz	35			
Carrots, whole (100)	6 #10			4 pc	25			
Corn, cream-style	6 #10			3 oz	35			
Corn, whole-kernel	6 #10			3 oz	35			
Onions, small whole (70–100)	6 #10			3 or 4	25			
Peas, June, 3 sieve	6 #10			3 oz	35			
Potatoes, white, tiny (160 or over)	6 #10			5 pc	32			
Potatoes, sweet, small (20–25)	12 #5			2 pc	12			
Sauerkraut	6 #10			3 oz	33			
Spinach, leaf	6 #10			3 oz	25			
Tomatoes, whole in puree	6 #10			4 oz	25			
Tomatoes, solid pack	6 #10			4 oz	25			
Tomatoes, puree, (1.05 specific gravity)	6 #10			2 oz	50			
Vegetables, mixed	6 #10			3 oz	35			
Yams, cut, vacuum pack	12 #5			2 or 3	12			
TOTAL								

FORM F

Date:_____

PRICE QUOTATION SHEET

Item	Specification	ABC Meats	Dixon's	G & H Supply	Metropolitan	Palmer	Wilson

Prices

FORM G

PURCHASE RECORD

Date	ABC Meat	Dixon's	Hawkeye	Mr. Gus	Monarch	Prairie Farm	Reed's	Wilson

Supplier or vendor

FORM H

PURCHASE RECORD

Date	Vendor	Total	Meat	Dairy	Groceries	Produce	Paper	Cleaning

FORM I

MEAL CENSUS

Day	Resident meals — General				Resident meals — Modified				Nourishment meal eq.*	Total	Nonresident meals — Employee			Nonresident meals — Guests			Total	total
	B	L	D	Total	B	L	D	Total			B	L	D	B	L	D		
1																		
2																		
3																		
4																		
5																		
6																		
7																		
8																		
9																		
10																		
11																		
12																		
13																		
14																		
15																		
16																		
17																		
18																		
19																		
20																		
21																		
22																		
23																		
24																		
25																		
26																		
27																		
28																		
29																		
30																		
31																		
Total																		

*Divide substantial nourishment (sandwich and beverage) by 3 and divide beverages by 5 to determine nourishment meal equivalents.

FORM J

PORTION COST

Recipe item_____

Number of servings_____

Cost per serving_____ Date_____

Ingredient	Amount	Unit price	Cost
Total			

Note: Multiply the amount of each ingredient times the unit price to determine the cost of each ingredient. Total these amounts to find the cost of the recipe. Divide this total cost by the number of servings to find the cost per serving.

FORM K

MONTHLY SUMMARY

Meals served

    Resident   _____

    Nonresident   _____

    Total   _____

    Man-minutes/meal   _____

                         _____

Total food expense   _____

    Average cost per meal   _____

Total nonfood expense   _____

    Average cost per meal   _____

Total labor expense   _____

    Average cost per meal   _____

Total equipment expense   _____

Total operating expense   _____

Total expense   _____

Total revenue   _____

Net dietary department expense   _____

3 BEEF

IN THE UNITED STATES, meat accounts for over 37 percent of the food budget. Knowledge of meat grades and cuts helps the buyer intelligently evaluate the alternatives available, whether selecting beef, pork, lamb, poultry, or fish.

WHOLESOMENESS

All animals are inspected for wholesomeness. The federal Meat Inspection Act requires all plants slaughtering meat or manufacturing meat products for interstate sale and shipment to operate under federal inspection. Federal inspection is carried out by the Meat Inspection Division of the Consumer and Marketing Service of the United States Department of Agriculture (USDA). The inspection for wholesomeness is done by veterinarians of the Meat Inspection Division or persons working under their supervision. Only meat that has first been inspected for wholesomeness may be graded. This inspection assures that the meat has come from a healthy animal and that it was processed in a sanitary plant.

IRRADIATION

To further assure that the meat supply is safe, purchase meat that has been irradiated to destroy the bacteria that cause food-borne illnesses and to lengthen shelf life. If irradiated meat, which is known to be safe, is not available, request the use of freshness indicators that change colors when the food is past its prime. The future of such indicators includes built-in thermometers that audibly alert operators to temperature changes and packages that self-destruct upon contact with microbial growth.

QUALITY GRADE AND CUT

Two important factors to consider when buying beef are the quality grade and the cut. The quality of each cut of meat will vary according to the grade; the higher the USDA grade, the more tender, juicy, and flavorful the meat will be.

Participation in the USDA grading program is voluntary. Federal law does not require that a food processor or distributor use the grade standards, but the standards are widely used because they are a dependable guide to quality for the consumer. Meat grading is provided for a fee by the USDA Agricultural Marketing Service to packers or others who request it.

Beef varies in quality more than any other kind of meat. Beef quality is judged by USDA inspectors and registered on the meat by rolling a ribbonlike imprint along the length of the carcass and across both shoulders. The imprint is a purple shield-shaped mark containing the letters *USDA* and the grade—Prime, Choice, Select, Standard, Commercial, Utility, Cutter, and Canner. When the carcass is divided into retail cuts, one or more of the grade marks will appear on each cut of meat.

The USDA grades are described as follows: Prime grade is the most tender, juicy, and flavorful. It is also the most expensive. Choice grade is also of high quality. It is moderately tender, juicy, and flavorful and the most widely available. It represents acceptable quality at a reasonable price. Select grade is not as juicy and flavorful as the first two because of the lack of fat marbling in the muscle, but it may be fairly tender. Standard and lower grades are seldom seen at the retail level. They are mature beef, with fat that will tend to be yellow rather than white, and will not be as tender as the other three grades. They are boned and used in ground beef.

COLOR AND AROMA

The color of meat is affected by advancing maturity. The older the animal, the deeper and darker the color. All meat is affected somewhat by oxidation and bacterial growth. Fresh veal is normally grayish pink, ranging from light to dark as the animal reaches maximum maturity. Fresh calf (carcasses that exceed maximum maturity for veal yet are too young to be classified as beef) has a normal color range of light grayish red to a moderate red. Fresh beef ranges from light grayish red to a very dark red.

Shady or blackish red beef, called "dark cutter," is believed to be the result of a reduced sugar content of the lean caused by shock or extreme excitement at the time of slaughter. The color of affected meat varies from slightly noticeable or two-toned to black with a gummy texture. The color does not affect the palatability. However, it does affect the acceptability and causes meat to be downgraded as much as one whole grade; consequently, dark cutter is seldom sold.

The aroma of fresh raw beef is reminiscent of commercial lactic acid. The odor is stronger in more mature animals. Meat from mature uncastrated male animals is often said to have an ammonia odor.

The process of aging beef enhances tenderness by breaking down tissues with the enzymes that are naturally present in the meat. Beef may be dry-aged (in open air) or wet-aged (in vacuum packaging). Optimum aging occurs at temperatures below 2°C (36°F) for 10–30 days. Prolonged storage may cause development of proteolytic or putrid odors from protein decomposition. Sour or tainted odors may occur from development of microbial growth, and rancid odors come from fat oxidation.

CARCASS VERSUS CUTS OF BEEF

The best way for an individual operation to decide whether to purchase carcass beef, wholesale cuts, or preportioned meat is to compare prices and quality of available products. It is possible that acquiring a combination of the three will be the best decision.

One of the most important requirements for carcass beef is careful cutting by an experienced butcher. If the cutting is to be done within the foodservice operation, special equipment and additional refrigeration will be needed. As a general rule for making weight estimates, count on 25 percent waste, 25 percent ground beef and/or stew meat, 25 percent steaks, and 25 percent roasts. This is a rough guideline to assist in making cost comparisons.

Yield grades are important to the buyer, for they identify the amount of usable meat on the carcass. Several beef carcasses all of the same quality grade can differ greatly in value because of differences in yield of usable meat. To identify these differences, the USDA has developed a system of five yield grades that are described by numbers 1 through 5 (Table 3.1). Grade 5 identifies the lowest yield of trimmed cuts per carcass. Grade 1 identifies the highest yield and is in short supply. Grades 2 and 3 are more available than Grade 1 and so are less expensive. Table 3.1 shows the percentage of trimmed cuts and waste for the five yield grades.

For a more exact comparison of carcass beef to preportioned cuts, first determine the cost of the carcass, adjusted for yield. To do this, divide the price per pound by the yield grade percentage to find the actual cost of usable meat per pound. For example, a side of choice beef weighing 300# before trimming, with a yield grade of 3, at a $1.60 per pound, would have an actual cost per pound of $2.20 or a total cost of $480.

$$\frac{\text{Price/pound}}{\text{Yield grade \%}} = \text{actual cost/pound or } \frac{\$1.60}{.728} = \$2.20$$

To compare this cost with the cost of any equivalent amount of preportioned cuts, multiply the current cost of each item by the amount that would be available from a single carcass and total. To determine the better buy between carcass beef and preportioned cuts, compare the two total costs. Table 3.2 may be of assistance.

Table 3.1. USDA system of yield grades for meat

Yield grade	Trimmed cuts	Bone and shrinkage	Fat trim
U.S. No. 1	82.0%	10.4%	7.6%
U.S. No. 2	77.4	9.9	12.7
U.S. No. 3	72.8	9.4	17.8
U.S. No. 4	68.2	8.9	22.9
U.S. No. 5	63.6	8.4	28.0

Table 3.2. Cost comparison for preportioned cuts with a single carcass

Preportioned cut	Approximate % of weight	Weight	Price/pound	Total value
Steaks				
Round	11	33	× _____ =	_____
Club and T-bone	5	15	× _____ =	_____
Sirloin	8	24	× _____ =	_____
Roasts				
Boneless rump	3	9	× _____ =	_____
Rib	6	18	× _____ =	_____
Chuck blade	9	27	× _____ =	_____
Chuck arm	6	18	× _____ =	_____
Ground beef, etc.				
Ground beef	11	33	× _____ =	_____
Stew meat	11	33	× _____ =	_____
Boneless brisket	3	9	× _____ =	_____
Waste	27	81	× –0– =	–0–
Total	100	300	=	_____

Three advantages of using preportioned cuts are (1) the amount purchased can be matched exactly to the amount needed according to the menu rather than adjusting the menu to match the available meat; (2) the buyer controls the quality, amount, and cost of the meat purchased; and (3) portions may be ordered to exact specifications so that each portion is the desired size. Extra portions may be kept on hand for changes in census, quick substitutions, or special orders. All these advantages result in greater control by the buyer.

INSTITUTIONAL MEAT PURCHASE SPECIFICATIONS

Beef has many descriptive names that are regional in nature. To provide a standard terminology and to simplify ordering, the USDA has written a set of Institutional Meat Purchase Specifications (IMPS), which describe each cut, specify the trim, and give an identification number. Meat can be ordered by the IMPS number, and the meat purveyor will know exactly what is desired. Under IMPS the order may be placed by thickness or by weight (not both) and by grade.

A free set of IMPS for beef, veal, lamb, sausage, and other cured and processed meats is available from the Meat Standardization and

Review Branch, Livestock, Meat, Grain, and Seed Division. AMS (Agriculture Marketing Service), USDA, Washington, DC 20250. A guide that is illustrated and easier to understand is "Meat Buyer's Guide," sold by the National Association of Meat Purveyors, 8365-B Greensboro Drive, McLean, VA 22102.

STEAKS AND ROASTS

Regardless of their quality grade, some cuts of beef are more tender than others. Cuts from the less-used muscles along the back will be more tender than those from the active muscles. The most tender cuts make up only a small proportion of the beef carcass; they are in greatest demand and are higher priced than others. These are the roasts and steaks from the rib and loin sections of the animal. These cuts can be prepared by use of dry heat and can be cooked to the desired degree of doneness without sacrificing tenderness.

The most tender, juicy, and flavorful steaks are porterhouse, T-bone, and club. Porterhouse has a generous portion of tenderloin along with the sirloin strip. T-bone steak is very similar to porterhouse except that it has a smaller amount of tenderloin. Club steak has the same large muscle as the porterhouse and T-bone but no tenderloin. Cuts from the rib section are roasts when they are two or more ribs thick. Sirloin steaks with the long flat bone are slightly more tender than those with the shorter bone. They have a larger portion of tenderloin and are sometimes called "top sirloin." The steaks with the small round or wedge-shaped bone have less bone, less fat, and less tenderloin and are less tender.

The less tender cuts make up the remainder of the beef carcass; they are slightly less desirable and less costly but are just as important as sources of nutrients. These are the roasts, steaks, stew meat, and ground beef from the shoulder, flank, and round. These cuts require tenderizing by heat and moisture during the cooking process, mechanically before cooking, or chemically by the addition of other ingredients during the cooking process. Names given to cuts of meat vary with the processor and the part of the country. The best guide for identifying beef (and veal) cuts is the standard terminology shown in the beef (and veal) charts at the end of the chapter.

Round steaks and roasts are among the leanest cuts and have little marbling. They have little or no bone. A full round steak contains three sections that vary in tenderness. Top round is the most tender and, if one of the top grades, may be broiled. Top round in other grades should be braised. Bottom round and eye of round steaks are less tender. The three different sections can be used as roasts. Sirloin tip roasts and steaks are similar in quality to top round and have no bone. Heel of round roast is boneless and very lean but is less tender. It should be pot roasted in all grades. Rump roast is tasty and fairly tender; it has a lot of bone which is often removed to make it easier to carve and serve. It can be oven roasted in the top grades but should be pot roasted in lower grades. Sirloin tip roast or knuckle is boneless with very little waste. It has good flavor but is not tender. Sirloin tip steak is cut from the roast. Neither the roast nor the steak is as tender as the regular sirloin.

GROUND BEEF

Once described by names such as "ground round" or "chuck," ground beef is sometimes called "hamburger," a German term for meat which is ground or pulverized. It is made by combining lean beef from lower grades with fat or trimmings from higher grades.

Different kinds of ground beef vary based on the cuts from which they are made and the ratio of lean to fat, and this is reflected in the price. Types of ground beef available are labeled as ground beef, ground chuck, ground round, and ground sirloin. The least expensive, which also contains the most fat, may be a good buy for casseroles, meat loaf, chili, and spaghetti.

Ground beef packaged under federal or state inspection at the packing plant or retail warehouse must contain no more than 30 percent fat. There are no federal regulatory specifications for fat content of ground beef produced in the retail store, although some states and local agencies have their own controls.

Leaner ground beef has less fat, usually no more than 20 percent, and more lean beef; thus, there is less shrinkage. This type of meat may be used for beefburgers, hamburger patties, and Salisbury steaks.

Ground beef that is quite lean has fat closely trimmed to no more than 10 percent. It is good for dieters because of the low fat content but tends to be less tender and juicy.

Ground beef may have vegetable protein product (VPP) added to it. VPP, formerly called "textured vegetable protein" (TVP), is a processed soy meal product (see Ch. 14). It is added to decrease cost without greatly decreasing the amount of available protein. However, it does change the amino acid ratio. Another advantage is that VPP absorbs moisture—both meat juices and fat. Since the flavor of meat is largely in the fat, this may produce a product with more flavor; however, for those attempting to reduce fat intake, this may be a disadvantage.

FROZEN MEAT

Fresh ground beef should be used within two days of delivery, or it should be frozen immediately. Meat that is delivered fresh and then frozen should be packaged in the proper size for the intended use, and the wrapping should be airtight. Ground beef may be stored two to three months with little loss of quality. Other cuts may be kept fresh three or four days or stored in the freezer for four to six months.

Frozen meat should be kept properly packaged and frozen at $-20°C$ (0°F) or below. The amount of available freezer space will influence your buying habits. Meats purchased frozen should be examined at the time of delivery for signs of thawing. Do not accept frozen meat that is partially thawed unless it is to be used immediately. Meat darkens on the surface when it is frozen in airtight wrapping.

Individually quick frozen (IQF) meats are available in the portion size desired. This process allows the number of portions needed to be thawed and the rest to remain frozen. When preportioned meats are frozen in a solid block, the whole package must be thawed for use. To avoid waste, buy block-frozen meat only if the entire package is needed for the intended use.

VARIETY MEATS

Variety meats are the internal organs of meat animals that are used for food. They include liver, heart, kidney, tongue, tripe, sweetbreads, and brains. Occasionally oxtail is grouped with variety meats. Brains, sweetbreads, and tripe are colorless and very delicate in flavor. The others are both colorful and flavorful as well as nutritious.

RESTRUCTURED MEATS

Restructured meat products are generally made from flaked, ground, or sectioned meat that is then shaped into roasts, steaks, chops, strips, or cubes. The process of restructuring involves reduction or modification of particle size, blending with a binder, and reforming into the desired product size and shape. The binders most commonly used are carageenen, which is a complex carbohydrate derived from red seaweed; oat bran, which tastes the same as carageenen; and soy protein, which is an inexpensive extender that has been used many years. Research and development have eliminated the off-taste formerly associated with this soybean derivative.

RETAIL CUTS OF BEEF

WHERE THEY COME FROM AND HOW TO COOK THEM

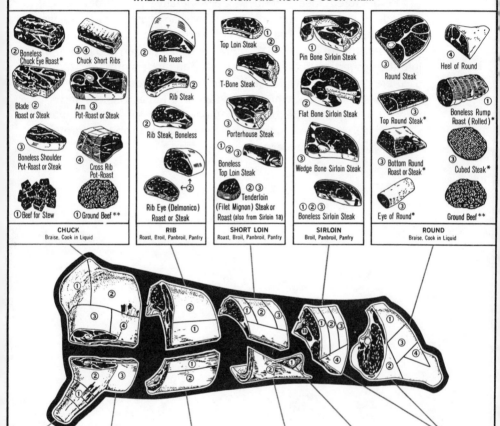

CHUCK
Braise, Cook in Liquid

- ② Boneless Chuck Eye Roast*
- ③④ Chuck Short Ribs
- Blade ② Roast or Steak
- Arm ③ Pot-Roast or Steak
- ③ Boneless Shoulder Pot-Roast or Steak
- ④ Cross Rib Pot-Roast
- ① Beef for Stew
- ① Ground Beef **

RIB
Roast, Broil, Panbroil, Panfry

- ② Rib Roast
- ② Rib Steak
- ② Rib Steak, Boneless
- ② Rib Eye (Delmonico) Roast or Steak

SHORT LOIN
Roast, Broil, Panbroil, Panfry

- ② ③ Top Loin Steak
- ② T-Bone Steak
- ③ Porterhouse Steak
- ① ② ③ Boneless Top Loin Steak
- ② ③ Tenderloin (Filet Mignon) Steak or Roast (also from Sirloin 1a)

SIRLOIN
Broil, Panbroil, Panfry

- ① Pin Bone Sirloin Steak
- ② Flat Bone Sirloin Steak
- ③ Wedge Bone Sirloin Steak
- ① ② ③ Boneless Sirloin Steak

ROUND
Braise, Cook in Liquid

- ③ Round Steak
- ④ Heel of Round
- ③ Top Round Steak*
- ① Boneless Rump Roast (Rolled)*
- ③ Bottom Round Roast or Steak*
- ③ Cubed Steak*
- ③ Eye of Round*
- ③ Ground Beef **

FORE SHANK
Braise, Cook in Liquid

- ① Shank Cross Cuts
- ① Beef for Stew (also from other cuts)

BRISKET
Braise, Cook in Liquid

- ③ Fresh Brisket
- Corned Brisket

SHORT PLATE
Braise, Cook in Liquid

- ① Short Ribs
- ① ② Skirt Steak Rolls*
- ① ② Beef for Stew (also from other cuts)
- Ground Beef **

FLANK
Braise, Cook in Liquid

- Ground Beef **
- Flank Steak*
- Beef Patties **
- ① Flank Steak Rolls*

TIP
Braise

- ④ ② Tip Steak*
- ④ ② Tip Roast*
- ④ ② Tip Kabobs*

*May be Roasted, Broiled, Panbroiled or Panfried from high quality beef.
**May be Roasted, (Baked), Broiled, Panbroiled or Panfried.

This chart approved by
National Live Stock and Meat Board

© National Live Stock and Meat Board

32

RETAIL CUTS OF VEAL

WHERE THEY COME FROM AND HOW TO COOK THEM

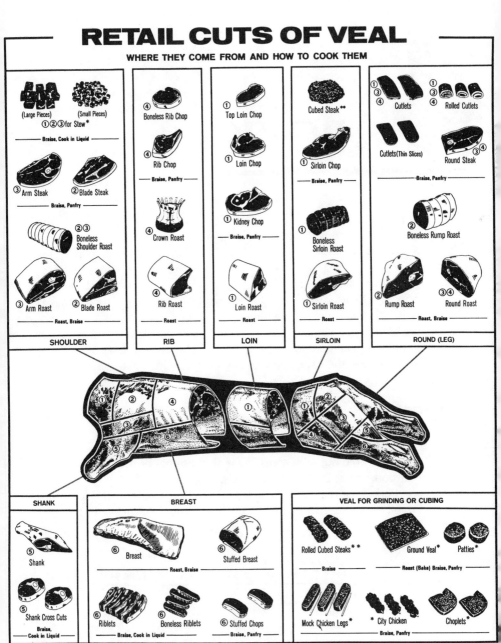

SHOULDER

(Large Pieces) (Small Pieces)
①②③ for Stew *

— Braise, Cook in Liquid —

③ Arm Steak   ② Blade Steak

— Braise, Panfry —

②③ Boneless Shoulder Roast

③ Arm Roast   ② Blade Roast

— Roast, Braise —

RIB

④ Boneless Rib Chop

④ Rib Chop

— Braise, Panfry —

④ Crown Roast

④ Rib Roast

— Roast —

LOIN

① Top Loin Chop

① Loin Chop

— Braise, Panfry —

① Kidney Chop

— Braise, Panfry —

① Loin Roast

— Roast —

SIRLOIN

Cubed Steak **

① Sirloin Chop

— Braise, Panfry —

① Boneless Sirloin Roast

① Sirloin Roast

— Roast —

ROUND (LEG)

① ③ ④ Cutlets   ① ③ ④ Rolled Cutlets

Cutlets (Thin Slices)   ③④ Round Steak

— Braise, Panfry —

② Boneless Rump Roast

② Rump Roast   ③④ Round Roast

— Roast, Braise —

SHANK

⑤ Shank

⑤ Shank Cross Cuts

— Braise, Cook in Liquid —

BREAST

⑥ Breast

⑥ Stuffed Breast

— Roast, Braise —

⑥ Riblets   ⑥ Boneless Riblets   ⑥ Stuffed Chops

— Braise, Cook in Liquid —   — Braise, Panfry —

VEAL FOR GRINDING OR CUBING

Rolled Cubed Steaks **   Ground Veal *   Patties *

— Braise —   — Roast (Bake) Braise, Panfry —

Mock Chicken Legs *   * City Chicken   Choplets *

— Braise, Panfry —

*Veal for stew or grinding may be made from any cut.

**Cubed steaks may be made from any thick solid piece of boneless veal.

This chart approved by
National Live Stock and Meat Board

© National Live Stock and Meat Board

33

4 PORK

PORK IS AN EXCELLENT SOURCE of protein, containing all the essential amino acids, and is not high in fat. It is high in iron, zinc, and the B vitamins, especially thiamine. Every pork cut is tender. It is one of the most usable and completely digestible foods. Pork is economical and available fresh, cured, smoked, and canned. Like beef, pork may be irradiated, which destroys the trichina bacteria and makes pork safe from food-borne illness even when not thoroughly cooked.

Pork, like beef, is inspected for wholesomeness, but it is not graded for quality. It is graded for yield in the same way as beef. Table 4.1 indicates the expected yield by grade.

FRESH PORK

Select fresh pork that is firm and grayish pink with as little exterior fat as possible but with some internal marbling, or request in writing that pork meets these specifications.

Pork roast may be purchased bone-in or boneless. The loin may be left whole or divided into blade end, center, and sirloin end roasts. The front quarter has a picnic or Boston butt and the hindquarter has

Table 4.1. USDA grade and expected yield for pork

Grade	Yield
U.S. No. 1	53% and over
U.S. No. 2	50 to 52.9%
U.S. No. 3	47 to 49.9%
U.S. No. 4	less than 47%

34

the fresh ham, which may be left whole or divided into the shank end and the rump. (See the pork chart provided at the end of the chapter.)

Chops include loin, rib, sirloin, and blade chops; butterfly and boneless chops; and chops with a pocket for stuffing. Steaks are either shoulder or cubed steaks. Pork ribs may be spareribs, back ribs, and country style. Hocks and feet, ground pork, fresh sausage, and variety meats such as liver, kidney, and heart complete the list of fresh pork items available.

CURED, SMOKED, AND CANNED PORK

Ham is the hind leg of pork. In addition to the fresh ham described in the preceding section, it may be purchased cured and canned or cured and smoked. Cured pork has been treated with curing ingredients. The salt is used for flavor and as a preservative. The nitrates and nitrites combine with the meat to develop the typical red color of the cured product. These ingredients are usually applied in a brine solution that may be pumped into the meat. This procedure permits curing to be done quickly. Most processors leave some water in the meat to retain juiciness and fine texture. In federally inspected plants if the curing results in an increase in weight of up to 10 percent, the ham must be labeled "water added." If the weight increases more than 10 percent, it must be labeled "imitation."

Cured ham is available ready to cook or fully cooked. The ready-to-cook ham should be baked to 70°C (160°F). The fully cooked is sometimes called "ready to eat" or "ready to serve," and no additional cooking is required. Ham is available with some, all, or none of the skin, fat, and bone removed. It may be purchased whole or divided into butt, center, and shank or into butt half and shank half. "Half" means that no center slices have been removed. Cured ham may be stored refrigerated in its original wrapper for up to two weeks. If properly wrapped, it may be stored in the freezer for up to 60 days, but frozen cured ham may lose some of its characteristic texture and flavor.

Country or dry-cured hams are produced by using a dry cure, slow smoking, and a long drying process. They are heavily salted and may require soaking and simmering before roasting. These may be called "Virginia" or "Georgia" hams depending on the state in which they are produced. Smithfield hams are hand-rubbed with salt, dry-cured in the salt, smoked, and dried. Because of their dryness and saltiness, these hams require soaking and simmering. Scotch hams are uncooked, boneless, mildly cured hams marketed in casings, a process that originated in Scotland. Prosciutto hams are Italian-type, ready to eat, highly seasoned, and dry. They are usually very thinly sliced and prepackaged or sold at delicatessen counters.

Canned ham usually consists of boneless cured ham pieces that are placed in a can, then vacuum sealed and fully cooked. A small amount of gelatin is added before sealing to thicken the natural ham juices. The traditional pear-shaped canned hams range in weight from 1 to 5 kilograms (1½# to 10#). Pullman canned hams are 10 centimeters (4 inches) square so that slices will fit on sandwich bread. Chopped ham is canned in oblong cans for retail sale and in pullman cans for wholesale purchasing. Unopened canned hams may be kept for several weeks. Read the label for refrigeration requirements.

A smoked picnic is not really a ham. It is from the shoulder, not the leg, and is usually less expensive and contains somewhat more internal fat and connective tissue.

Smoked loin roasts, smoked loin and rib chops, and Canadian-style bacon are cured and smoked cuts from the pork loin. Canadian-style bacon consists primarily of the boneless "eye" of the loin muscle that is defatted and shaped into a compact roll and cured. Ham shank rolls are very similar to Canadian-style bacon but are less expensive.

Bacon is another popular cured and smoked pork product. It is cut from the pork side and is sold regular sliced (35 to 45 slices per kilogram or 16 to 20 slices per pound); thick sliced (30 to 35 slices per kilogram or 14 to 18 slices per pound); or slab style, which is uncut.

Jowl, hocks, feet, and salt pork complete the list of cured and/or smoked pork items that are available.

Sausages are available in more than 200 variations. Pork is commonly used, but sausages may contain beef, veal, chicken, and other meats. The best source of information about each is the label. The label tells what the product is and the ingredients that are used, indicating the quantity by listing ingredients in decreasing order of presence and including other information, such as the preservatives and spices and seasonings used to give each product its distinctive flavor. In addition to salt, red, white, and black pepper; allspice; anise; bay leaves; cinnamon; and many other flavorings may be used. Sausages and ready-to-serve meats are generally grouped according to the processing method used.

Fresh sausages are made from selected cuts of fresh meats. They have been neither cooked nor cured, so they must be cooked thoroughly before serving. Fresh pork sausage is available as bulk, links, patties, or country style. Other varieties include kielbasa (Polish), Italian, bratwurst, chorizo (link or bulk), and thuringer. Some of these varieties are also available smoked, cured, and/or fully cooked.

Uncooked smoked sausage sometimes includes cured meat that has been smoked but not cooked, so it must be cooked thoroughly before serving. Included in this category are smoked pork sausage, kielbasa, mettwurst, and smoked country-style sausage.

Cooked sausages are usually made from fresh meats that are cured during processing and are fully cooked. They are ready to eat, but some may be served hot. Included in this group are blood sausage, blood and tongue sausage, bratwurst, kiszka, liver loaf, yachwurst, braunschweiger, and liver sausage.

Cooked smoked sausages are made from fresh meats that are cured during processing and are fully cooked and ready to eat. Included in this group are bierwurst (beef salami), bologna, mettwurst, cotto salami, and frankfurters (wieners, German-style mortadella, and prasky). Also included are kielbasa, Krakow, New England sausage (Berliner), smoked thuringer links, teawurst, and Vienna sausage.

Dry and/or semidry sausages are made from fresh meats that are cured during processing and are either smoked or unsmoked. They are prepared by carefully controlled bacterial fermentation that acts as a preservative and develops the flavor. Most dry sausages are salamis. Most semidry sausages are of the summer sausage type. Both are ready to eat and should be refrigerated. Included in this group are summer sausage, cervelat (farmer), thuringer, salami, Genoa, German (hard), kosher, Milano, chorizo, frizzes, Lebanon bologna, Lyons, medwurst (Swedish), Metz, mortadella, and pepperoni.

Specialty meats (luncheon meats) are made from fresh meats that are cured during processing. They are fully cooked, sometimes baked. The loaves included in this group are Dutch, ham and cheese, honey, jellied tongue, minced ham, old-fashioned, olive, pepper, pickle and pimiento, chopped ham, headcheese, minced roll, scrapple sylta, and Vienna sausage.

RETAIL CUTS OF PORK

WHERE THEY COME FROM AND HOW TO COOK THEM

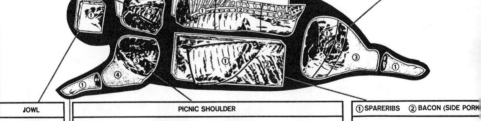

*May be made from Boston Shoulder, Picnic Shoulder, Loin or Leg.

This chart approved by
National Live Stock and Meat Board

© National Live Stock and Meat Board

5 LAMB

LAMB IS TASTY AND NUTRITIOUS. The younger the animal when it is brought to market, the leaner and more tender its meat. Hothouse lamb comes from milk-fed sheep up to two months old. Spring lamb comes from sheep that have been weaned and are from three to five months old. Most lamb is from sheep between five and seven months of age and is known for its tenderness and delicate flavor. Meat from sheep older than a year has a stronger taste and is less tender. The meat is red and has cream-colored fat and large white bones. About 92 percent of the lamb sold in the United States is from sheep raised domestically.

QUALITY GRADE

Each USDA lamb grade is a measure of a distinct level of quality. There are five grades. Prime, Choice, and Good are similar to the same grades in beef. Utility and Cull are seldom sold as retail cuts. Prime is the most tender, juicy, and flavorful. Choice is the grade produced in the greatest volume, and suppliers have found that this quality satisfies most consumer needs.

In addition to lamb, the USDA has grades for yearling mutton, which is meat from sheep one to two years old, and for mutton, which is meat from older animals.

Grading of lamb is provided by the USDA Food and Quality Service to meat packers and others who request and pay for it. Approximately two-thirds of all lamb produced is graded for quality.

YIELD GRADE

Lamb carcasses may also be graded for yield of trimmed retail cuts. (See yield grade table in Ch. 3.) Although only a small amount of lamb is graded for yield, the service is available, and the yield grade shield may appear on larger wholesale cuts. Imported lamb is not graded for quality or yield by the USDA, since it is usually imported frozen and only fresh meat can be graded. However, imported lamb must be inspected for wholesomeness by a system approved by the USDA and must bear the inspection legend of the exporting country.

CUTS

Lamb is available year-round. Most people are familiar with legs and chops. Many other cuts are equally good, and some are more economical. (See the lamb chart at the end of the chapter.)

Leg of lamb is cut in a variety of ways—American-style, French-style, boneless, rolled, or butterflied. It may be whole with the sirloin portion left on, or it may be cut as sirloin half or shank half roast, center leg roast, or leg steaks. Chops and roasts, either boneless or bone-in, may be cut from the sirloin section. It costs less to buy the whole leg than the sirloin and shank halves separately.

Lamb loin may be cut as T-bone or boneless chops, or it may be left as a roast. Loin and rib chops are the most expensive chops, followed by sirloin chops, then shoulder chops.

Lamb shoulder is usually less expensive than the leg. The roast may be purchased square-cut; presliced and tied; or boned, rolled, and tied. Arm and blade-bone shoulder chops may be cut from this roast. The meat may also be used in stews, in curries, and as brochettes.

Shanks may be left whole or cut into sections for stew.

Ground lamb may be used for lamb patties or casseroles or mixed with other meats for meat loaf. It is made from the neck, breast, or shank.

STORAGE

Lamb that is loosely wrapped may be refrigerated at 0° to 1°C (32° to 34°F) for up to three days. Lower temperatures are required for longer storage times.

RETAIL CUTS OF LAMB

WHERE THEY COME FROM AND HOW TO COOK THEM

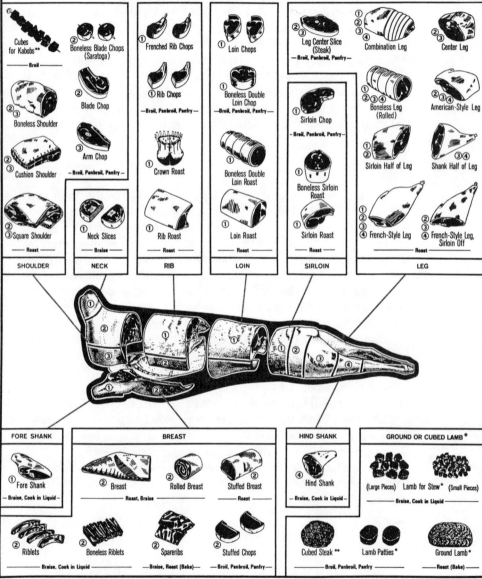

Cubes for Kabobs**
— Broil —

② Boneless Blade Chops (Saratoga)

② Blade Chop

③ Arm Chop
— Broil, Panbroil, Panfry —

② ③ Boneless Shoulder

② ③ Cushion Shoulder

② ③ Square Shoulder
— Roast —

SHOULDER

① Frenched Rib Chops

① Rib Chops
— Broil, Panbroil, Panfry —

Crown Roast

Rib Roast
— Roast —

② Neck Slices
— Braise —

NECK

RIB

① Loin Chops

① Boneless Double Loin Chop
— Broil, Panbroil, Panfry —

① Boneless Double Loin Roast

① Loin Roast
— Roast —

LOIN

② Leg Center Slice (Steak)
— Broil, Panbroil, Panfry —

① Sirloin Chop
— Broil, Panbroil, Panfry —

① Boneless Sirloin Roast

① Sirloin Roast
— Roast —

SIRLOIN

② ③ Combination Leg

② ③ Center Leg

② ③ ④ Boneless Leg (Rolled)

② ③ ④ American-Style Leg

① ② Sirloin Half of Leg

③ ④ Shank Half of Leg

① ② ③ ④ French-Style Leg

② ③ ④ French-Style Leg, Sirloin Off
— Roast —

LEG

FORE SHANK

① Fore Shank
— Braise, Cook in Liquid —

BREAST

② Breast
Rolled Breast
② Stuffed Breast
— Roast, Braise — — Roast —

② Riblets
② Boneless Riblets
— Braise, Cook in Liquid —

② Spareribs
— Braise, Roast (Bake) —

Stuffed Chops
— Broil, Panbroil, Panfry —

HIND SHANK

④ Hind Shank
— Braise, Cook in Liquid —

GROUND OR CUBED LAMB*

(Large Pieces) Lamb for Stew* (Small Pieces)
— Braise, Cook in Liquid —

Cubed Steak**
— Broil, Panbroil, Panfry —

Lamb Patties*

Ground Lamb*
— Roast (Bake) —

* Lamb for stew or grinding may be made from any cut.
**Kabobs or cubed steaks may be made from any thick solid piece of boneless Lamb.

This chart approved by
National Live Stock and Meat Board

© National Live Stock and Meat Board

6 SEAFOOD

FRESH FISH HAS A SWEET, fresh, mild, pleasant smell, not a strong "fishy" odor. It has clear, bright, bulging eyes and shiny scales that adhere compactly to the skin. The gills are red and free from slime. Purchase fresh fish that has been packed in ice. Test for freshness by pressing gently with the finger; if the impression remains in the flesh, the fish is stale. Freshness can be assured if the seafood has been irradiated, a process that kills the bacteria and parasites that infect fish and shellfish and delays spoilage.

Seafood purchased in a market is inspected by the Food and Drug Administration (FDA), so although there has been concern about contaminants in water affecting the fish sold, the incidence of illness is very low. To reduce the risk, buy fresh fish, wash it thoroughly, use it soon, cook to well done, and remove the skin and fat, where the bacteria collect. Don't store cooked seafood near raw fish because cross-contamination occurs very easily.

Fish is extremely perishable and should be used as soon as possible. Wrap in foil or place in a pan and cover with crushed ice. Even when properly stored, fish will start to decay within three days. For long storage, freezing is best.

Frozen fish and shellfish are available the year around. Most fish that is available fresh is available frozen, most likely as fillets or steaks. Packages should be kept frozen solid in a tight moistureproof, vaporproof wrapping. An ice glaze is often used to protect seafood from drying out. Any discoloration may indicate a deterioration of quality.

To defrost fish, place the wrapped fish in the refrigerator. If thawed at room temperature or in warm water, fish becomes soggy. If it is to be broiled, baked, poached, panfried, or deep-fat fried, the fish does not have to be defrosted before preparation.

FORMS AND CUTS

Fish is marketed in various forms and cuts.

Whole fish is sold just as it comes from the water.

Drawn fish has had the fins, scales, and entrails removed.

Dressed fish has also had the head and tail removed.

Fillets of fish are the sides cut lengthwise away from the backbone. Single fillets are cut from one side of the fish; butterfly fillets are the two sides of the fish held together by the uncut flesh and skin of the belly. Fillets are usually skinless and boneless.

Steaks are cross-section slices from large dressed fish cut 1.5 centimeters or 0.6 inch thick to 2.5 centimeters or 1 inch thick. A cross section of the backbone is the only bone in a steak.

Chunks are cross sections of large dressed fish.

Fish portions are made of less expensive and less desirable varieties such as whitefish. They are available buttered, portioned and breaded, or portioned, breaded, and cooked. Fish portions are made up by cutting them from frozen fish blocks then coating them with butter, packaging, and freezing; coating them with butter or oil, breading, packaging, and freezing; or coating them with butter or oil, breading, cooking, packaging, and freezing. They may be square, oblong, or fillet shaped. United States Department of Interior (USDI) standards require that raw portions must contain not less than 75 percent fish flesh and fried portions not less than 65 percent fish flesh.

Fish sticks are cut from frozen fish blocks, coated with oil, breaded, partially cooked, packaged, and frozen. USDI standards require that they contain not less than 60 percent fish flesh.

Canned fish are most commonly tuna or salmon, but sardines, mackerel, anchovies, shrimp, crabmeat, or clams may also be canned. Canned salmon may be any of the varieties used as fresh fish. Canned tuna may be fancy or solid pack, chunk-style, flaked, or grated. It may be salted and seasoned, reduced salt, or salt-free and may be packed in oil or water.

The label will tell the grade, size, species, contents, weight, and ingredients added. Canned fish has a shelf life of two to three years. It has a 100 percent yield, so it is easy to determine the amount to purchase if the serving size and number to be served is known.

Smoked fish is becoming increasingly popular in restaurants.

LEAN FISH

There are a variety of ways fish may be classified. One is by the amount of fat because of the relationship of fat to other nutrients. Fatty fish provides twice as much protein, fat, iron, phosphorus, potassium, and niacin and other vitamins as well as twice as many calories as lean fish.

There are several popular varieties of lean fish.

Flatfish are an assortment of related fish including true flounder, fluke, plaice, sole, turbot, and halibut. They are mild tasting, low in calories, and available in fillets or whole for stuffing. Sole is closely related to flounder but is more expensive. Lemon sole has a firm texture and sweet taste. Gray and Dover sole are the elite of fish. Turbot is a European flatfish with good flavor. It also is closely related to flounder. Halibut is a large flounderlike fish sold as steaks. It is ideal for broiling or poaching and has a mild nonfishy flavor.

Cod is extremely versatile. It is easy to eat whole because its few bones are large and easy to remove. Smaller cod are called "scrod."

Hake and haddock are related to cod and are tasty; they are good for stew and chowder.

Pollock, also related to cod, has a similar taste but is darker.

Bass may be steamed or poached whole. Sea and striped bass may be baked.

Grouper is related to bass and comes from southern waters.

FATTY FISH

There are several popular varieties of fatty fish.

Bluefish is smooth textured and has a sweet taste when fresh. It may be grilled whole, or fillets may be broiled or fried.

Catfish that is farm grown has a mild flavor. It is broiled, fried, or grilled whole and is easy to eat because it does not have small bones in the flesh.

Salmon is available mainly from May to October; it combines a rosy pink color with a rich, buttery taste. Salmon is identified by variety, since there are differences in color, texture, and flavor. The most expensive varieties are deeper red and have a higher fat content. There is only one variety caught in the Atlantic. Pacific varieties, in order of expense, are sockeye (red), king or chinook, coho (silver or medium red), humpback (pink), and keta or chum. Usually available in steak form, they may be cooked by almost any method; they may also be poached and served cold or smoked.

Trout is related to salmon. Lake and river trout are gray fleshed, while sea trout is pink fleshed. They may be grilled whole or filleted and served with almonds.

Red snapper is a large red-scaled fish with a unique taste. Steaks and fillets are best when broiled in butter. When buying, specify Gulf rather than Pacific snapper.

Mackerel is available whole or cut into steaks. West Coast varieties are known as "bonito." Tuna (a member of the mackerel family) is available as steaks that have a hearty texture and meaty taste. It is either light or white. The albacore is the only white meat tuna of the species and is the most expensive. Light tuna may be skipjack, yellowfin, or bluefin.

Swordfish is an expensive novelty; the steaks are best when broiled in butter.

Monkfish is a newcomer to the market and differs from traditional fish. It has a skin that is removed in the processing plants before shipping. The fish are round and covered with a membrane that must be removed before cooking. Each fish weighs from ½ kilogram to 1½ kilograms or 1# to 3# and because of its flavor has been called "poor man's lobster."

Seafood may be grouped by color and flavor (Table 6.1). The different fish in each category taste somewhat alike and can be prepared in the same way. If a particular fish is unavailable, another from the same group may be used in its place.

SHELLFISH

Shrimp is available either fresh or frozen. Three main domestic varieties are northern, North Pacific, and southern. Peeled and deveined shrimp are purchased in 3# or 5# bags and are specified by the number of shrimp per pound. The smallest are popcorn shrimp and have 40 or more per pound. Small has 31 to 35 per pound; medium has 26 to 30 per pound and is the size most frequently used for salads. Large has 21 to 25 per pound; jumbo has 16 to 20 per pound and is the most frequently used size for shrimp cocktail and hors d'oeuvres. Extra jumbo has 10 to 15 per pound and is the most expensive. Cost per pound decreases as the number per pound increases.

Frozen raw breaded shrimp are made from whole, clean, headless shrimp that have been peeled and deveined; they may be round or fantail. The shrimp are coated with a wholesome batter and/or breading. Shrimp is called "whole" if it consists of five or more segments of shrimp flesh. According to U.S. standards, Regular Breaded is frozen raw breaded shrimp containing a minimum of 50 percent of shrimp material. Lightly Breaded contains a minimum of 65 percent of shrimp material. Using standards set by the National Marine Fisheries Service, breaded shrimp may be U.S. Grade A, U.S. Grade B, or Substandard. Breaded and headless shrimp in the shell are usually purchased in 3#, 5#, or 6# boxes. Shrimp shapes are a less expensive, fabricated product made from shrimp pieces that are extruded in the shape of shrimp and breaded. Shrimp patties are also available.

Table 6.1. Color and flavor categories for seafood

White Meat: Very Light, Delicate Flavor

Cod	Orange roughy	Rex sole
Dover sole	Pacific halibut	Summer flounder
Haddock	Pacific sanddab	Yellowtail flounder
Lake whitefish	Petrale sole	Yellowtail snapper

White Meat: Light to Moderate Flavor

American plaice/dab	Mahi Mahi	Starry flounder
Arrowtooth flounder	Pacific whiting	White king salmon
Butterfish	Red snapper	White sea trout
Catfish	Rock sole	Whiting
English sole	Snook	Winter flounder
Lingcod	Spotted sea trout	Wolffish

Light Meat: Very Light, Delicate Flavor

Alaska pollock	Grouper	Smelt
Brook trout	Pacific ocean trout	Walleye
Giant sea bass	Rainbow trout	White sea bass

Light Meat: Light to Moderate Flavor

Atlantic ocean perch	Lake chub	Pollock
Atlantic salmon	Lake herring	Pompano
Buffalofish	Lake sturgeon	Rockfish
Carp	Lake trout	Sablefish
Chum salmon	Mako shark	Sand shark
Crevalle jack	Monkfish	Scup/Porgie
Croaker	Mullet	Sheepshead
Greenland turbot	Northern pike	Silver (coho) salmon
Jewfish	Perch	Striped bass
King (chinook) salmon	Pink salmon	Swordfish

Scallops are the muscles removed from scallop shells and have a characteristic sweet flavor. They are packed fresh in 8-gallon containers, but smaller amounts may be available from the supplier. The two most common sizes of sea scallops are 20 to 30 pieces per gallon and 30 to 40 per gallon. Bay scallops are much smaller and are usually available only in areas close to the water from which they are harvested. Formed scallops, which are made of pieces and shaped to look like scallops, are available at a lower cost. They are usually sold fresh but are available frozen.

Lobsters purchased whole are shipped alive. The meat is white and sweet. Warm water varieties come from the Caribbean and have a firm shell. Cold water varieties come from South Africa, New Zealand, and Maine and have a harder shell. They range in size from the chicks, which are 1# or less, to jumbo, which are 2¾# to 3½# each with intermediate sizes of 1¼# to 1½#; selects at 1½# to 1¾#; and extra selects at 1¼# to 2#. They are shipped in crates of 25 lobsters each.

Lobster tails are usually purchased frozen. They are packed in 5# boxes and are sized within the box. The three most common sizes are 8 to 10 ounces each, 10 to 16 ounces, and 16 to 20 ounces. Slipper lobster, the small pieces of meat removed from the remainder of the lobster after the tail has been removed, is packed in 5# boxes and shipped frozen.

Crab is a seafood that includes many varieties. The most common ones used in foodservice are snow, Dungeness, blue, stone, and king. Snow crabs may be either opillio, which are small and have less quality, or baridi, which are larger and select. They are packed in 5# boxes. Dungeness crabs are harvested on the Pacific coast. They are round to oval and are packaged whole in 5# boxes, which may be shipped fresh or frozen. Blue crabs are small with soft shells. They are harvested along the Atlantic coast and are usually shipped fresh in 5# and 10# boxes. Stone crabs are ½# to ¼# each and are shipped in 10# and 20# bags. King crabs are named for the area where they are harvested. One of the more popular varieties is the Alaskan king, which comes from the waters off the Alaskan shores. King crabs are precooked in the processing plant, and the legs are removed and shipped frozen. The legs are not sized, so each box will have a variety of sizes. The crabmeat from the rest of the body is removed and packed in 5# boxes and shipped frozen. Whole crabs, crab legs, and crab meat are all considered a delicacy and are quite expensive. "Sea legs" are made from a lower-quality fish with a crablike texture that has been injected with a crab taste and dye and formed to resemble crab legs.

Clams are packed and shipped in bushel bags. The two most common sizes are cherrystone, which are 250 count, and littlenecks, which are 300 to 350 count. Fresh clams are available locally. Frozen clams, in the shell, are packed in 30# containers and are usually littlenecks. Smaller clams, called "steamers," are 400 count. They are shipped frozen in 3# bags. Clams are also available canned whole or minced.

Oysters are available and plentiful from September to April. They may be purchased in the shell, but most are shelled, packed in gallon containers, and shipped fresh. The sizes most often available are selects, which are 200 per gallon, and standards, which are 250 per gallon. West Coast frying oysters, 8 to 10 per pound, are shipped fresh and are more expensive because of the large size.

IMITATION SHELLFISH

Surimi is imitation crab, lobster, shrimp, and scallops. It is made by repeatedly washing mechanically deboned fresh fish with chilled water until it becomes odorless and colorless. The pulp is drained and strained before sugar, salt, potassium sorbate, sodium glutamate, and starch are added. It is shaped into flat rectangular blocks and frozen. This surimi paste can be manufactured into products that have an elastic and chewy texture similar to that of shellfish, such as seafood cakes, fish balls, fish sausages, and other seafood analogs.

7 POULTRY

POULTRY IS FEDERALLY INSPECTED for wholesomeness; then it is graded for quality. The wholesomeness inspection, or lack of it, is a major concern to consumers and government officials. Buyers must be assured that the products being purchased are safe to eat. One of 20 food-borne illnesses are caused by poultry. The contamination can occur at any point in the process. The only sure method of preventing this contamination is irradiation. Studies conducted by the USDA show that 99.5 to 99.99 percent of the salmonella in chicken is destroyed by the dose of irradiation prescribed by the FDA.

U.S. poultry grades apply to turkey, chicken, duck, goose, and guinea. The grade shield appears on the package of chilled or frozen ready-to-cook birds or parts. Grade A birds are fully fleshed, meaty, and free from defects such as pinfeathers and skin tears. Grade B birds may be less attractive in finish and appearance and slightly lacking in meatiness. A broken wing or broken skin may cause the product to be lowered to Grade B. A bird not displaying a grade marking is usually Grade B.

Poultry labels should be used to identify the type of product desired for your meals. Young poultry means tender poultry. Young chickens are labeled "fryer," "broiler," "roaster," "young chicken," "capon," or "Rock Cornish game hen." Young turkeys are labeled "young hen," "young tom," "young turkey," or "fryer-roaster." Older poultry is labeled as "mature," "yearling," "stewing," or "old." It is generally used for soups, stews, roasting, and casseroles.

Turkey and chicken are used almost interchangeably in many facilities. Turkeys are larger, thus the percentage of usable meat in relation to bone is greater than in chickens. Turkey pieces are larger and yield better slices for sandwiches.

Care should be used in handling poultry products. Poultry is very perishable. Frozen poultry should be kept hard frozen until time to thaw. Use fresh-chilled poultry within one to two days.

There are three ways to thaw a frozen bird. (1) In the refrigerator—place poultry, still in its original wrap, in a tray or pan. Thaw in the refrigerator for one or two days until pliable. Turkeys weighing 18# or more may take up to three days to thaw. (2) In cold water—if poultry needs to be thawed on short notice, it should be placed in cold water in its watertight wrapping. Change the water often to hasten thawing. Time required will range from one hour for small birds to six to eight hours for large turkeys. Thaw until pliable. (3) In the cooking process—take the bird directly from the freezer and place in the oven. Increase the cooking time by 10 to 15 percent. If poultry is not to be cooked immediately, place it in the refrigerator.

CHICKEN

Chickens are categorized by size and age. Rock Cornish game hens are immature chickens with the Cornish breed in their bloodlines. They are from 5 to 6 weeks old and weigh from 1# to 2#. Broilers are about 4 to 8 weeks old and weigh from 2½# to 4#. Fryers are about 7 to 13 weeks old and weigh from 3½# to 6#. About 90 percent of all chicken sold in the United States averages 3½#. Young roasters are more than 10 weeks of age and weigh 6# to 8#. Capons are 10 to 12 weeks old and weigh about 6# to 10#. These are castrated male birds that are noted for their generous quantity of tender white meat. Stewing hens are 12 to 18 weeks old and weigh 4½# to 6#.

The price per pound for poultry is often low, yet the price per serving is increased by the high percentage of bone. For example, a 5# stewing hen will yield about 21 ounces of cooked boneless meat.

When purchasing chicken, select the type of bird best suited for the intended cooking method. If dry heat will be used, purchase young hens, fryers, or broilers. If moist heat will be used, buy roosters or old hens.

Chickens are sold whole, cut up, and further processed. Whole chickens are often packed with the heart, gizzard, and liver inside the cavity. Cut up chickens are halved, quartered, or cut into eight or nine pieces. Further-processed chicken products include cooked, smoked, ground, boned, marinated, seasoned, breaded, and dried items. A sampling of products includes nuggets, patties, strips, and pot pies. Novelty items such as frankfurters, bologna, and corn dogs can be made from chicken. Because of the added ingredients and spices, they assume the characteristic flavors of the same products made of beef and pork. Chicken can be preprocessed into gourmet items such as chicken Kiev, chicken cordon bleu, and stuffed breast. These items provide a quick-to-prepare special menu item, but they are very expensive and so should be used with discretion.

Poultry may be purchased chilled or frozen. Chilled forms of packing include wet- and dry-ice packs. In both forms, the product is unfrozen and ready for immediate use. Wet-ice packed chickens are chilled and packed 20 to 24 birds per carton, then covered with ice chunks or shavings. Dry-ice packed chickens are similarly packed in cartons but covered with carbon dioxide "snow" instead of wet ice.

Frozen pieces of chicken are convenient and easy to use because exact sizes and kinds of pieces desired are available. Breasts, legs and thighs, and wings and drumettes are packed in separate boxes so that the exact kind and amount needed can be purchased. Partial boxes may be returned to the freezer, still frozen, for storage until needed.

Frozen chicken pieces are available prebreaded. They are usually packed partially cooked but are also available in heat-and-serve form. Other breaded specialty items include breast patties, breast fingers, nuggets, breast fillets, breast strips, white and dark patties, and formed patties with added textured vegetable protein.

Frozen pulled chicken and frozen diced chicken are not only natural chicken products but have been handled in such a way as to extend the shelf life as well as the safe holding time after incorporation into food items. The pieces are individually quick frozen (IQF) so that part of the container contents may be removed and used, and that remaining can be left frozen and returned to the freezer without damage to the product. Thawing is not necessary before cooking.

TURKEY

Turkey is available whole or in parts. Raw turkey parts include breasts, tenderloins, drumsticks, wings, thighs, drumettes, or cubes and may be purchased either fresh or frozen. The wise buyer will investigate all the options and select according to the intended use.

Whole turkeys may be young hen, young tom, young turkey, or fryer-roaster. They may be prebasted. The birds may weigh up to 40#, but those ranging from 18# to 22# are generally preferred. Large turkeys provide a better yield and usually are priced a few cents a pound less than smaller hens or toms. They may be roasted whole or may be cut into halves or quarters for easier handling. The cut up birds require less cooking time and less oven space. Cooked yields are difficult to estimate. From 38 to 52 percent or approximately 50 percent of the dressed weight of the bird is usable meat, depending on a number of factors: a hen has slightly better yield than a tom; larger birds have proportionately better yields; birds cooked at the suggested temperature of 175°C (325°F) will have a higher yield than those cooked at higher temperatures. Precooked boneless turkey parts will yield 92 percent of their "as purchased" weight. Ready-to-cook boneless turkey parts will yield 61 percent of their as purchased weight.

Turkey breasts may be whole or split, skin on or skinless, seasoned or unseasoned, formed or natural shape, or smoked.

Boneless turkey roasts may be raw or cooked; refrigerated or frozen; seasoned or unseasoned; breast, thigh, or mixed parts; or rolled, tied, and oven ready. Cooked rolls have very little shrinkage, and it is easy to estimate the number of portions they will yield.

Formed turkey rolls are available in your choice of all white meat, all dark meat, or a combination. The combination of white and dark meat usually costs less. The rolls are formed by forcing pieces of turkey, pureed turkey, or gelatin into a tube mold.

Ground turkey typically has 15 percent fat or less. It is purchased raw and may be unseasoned or seasoned as sausage.

Breaded turkey may be purchased as cutlets, patties, fingers, or nuggets and can be any proportion of dark and light. It is available fully cooked and frozen.

Precooked diced turkey is available in 5# bags. Because it is precooked and boneless, it has a yield of 100 percent. The pieces are individually frozen, so part of the package may be used and the rest returned to the freezer without product deterioration.

Cured deli items include turkey ham, pastrami, hot dogs, and bologna as well as smoked boneless turkey breasts. Turkey ham, a fully cooked and smoked turkey roast, has a flavor very much like traditional ham. It is usually available in a roast that is about 10#. It is less expensive than ham and has a yield of 100 percent because it is fully cooked and boneless. Pastrami, hot dogs, and bologna made from turkey are similar to those made with beef and combination beef and pork but are lower in fat and calories. It is necessary to compare price per pound or ounce of boneless usable meat when deciding what product to purchase. For example, to yield 100 2-ounce servings of turkey, three of the possible sources are a 13¾# precooked turkey roll, a 20½# ready-to-cook turkey roll, or 6½ 35-ounce cans of boned turkey.

DUCK

Duck is a staple of contemporary American cooking. The availability of convenience duck products means that operators who wish to serve a specific cut no longer have to find uses for the rest of the bird.

Duckling is available whole and in parts, usually boneless breasts or bone-in thigh and leg portions. Boneless breasts may be purchased marinated. Duck tenderloins are breaded for frying. Bulk packs of boneless and skinless duck meat are available; strips are cut from the breast, and chunks come from the thigh and leg.

White Peking ducks are the most-consumed variety. Barbarie and Muscovy ducks are larger and have a more-pronounced flavor. Whole ducklings are sold in ½# increments from 3# to 6#. White Peking boneless breasts are available in 4#, 6#, and 8-ounce portions, and Barbaries come in 8#, 12#, 16#, 20#, and 24-ounce sizes. Duck meat is packed in 5# bulk boxes.

OTHER POULTRY

Quails are available fresh or frozen and usually sold semiboneless with only the leg and wing bones intact. They usually weigh about 4 ounces and are sold 24 head to a case.

Pheasant is an all white meat bird with a delicate gamey flavor usually served during the fall hunting season.

Specialty birds such as guinea hen, squab, poussin (baby chicken from French breeding stock), goose, and partridge are seldom purchased and then only for fine dining.

8 EGGS

EGGS ARE AMONG the most versatile of foods. They are prepared in a variety of ways for breakfast and are an ingredient in salads, desserts, entrees, and baked products. They are an inexpensive and convenient source of high-quality complete protein.

The USDA shield provides the buyer with information about grade and size of eggs. The letter designation used for grade is not influenced by the egg size. Eggs are graded according to condition, interior quality, and appearance of the shell. The grading is done under controlled conditions and with the supervision of a USDA inspector.

Eggs may be graded AA or Fresh Fancy, A, and B. The USDA Quality Control Program sets up standards requiring Grade AA eggs to reach the market quickly under strictly controlled conditions, guaranteeing the consumer a fresh, top-quality product.

Grade AA eggs cover a small area when broken; the white is thick and stands high, and the yolk is firm and high. Grade A eggs cover a moderate area when broken; the white is reasonably thick and stands fairly high, and the yolk is firm and high. Grade B eggs cover a large area when broken; the white is thin, and the yolk covers a larger area. When appearance is important, as in poaching or frying, Grade AA and A eggs are preferred. When appearance is not important, as in general cooking and preparation of baked products, Grade B eggs are satisfactory.

STORAGE AND HANDLING

Proper storage and handling are important in maintaining quality because Grade AA eggs can rapidly lose quality to become Grade B.

Eggs kept at room temperature may lose more quality in one day than in one week under refrigeration. Kept at 5°–10°C (40°–45°F), eggs will retain their quality for several weeks.

Shell eggs should be delivered in refrigerated trucks or be cooled to less than 15°C (60°F) during transportation and not be allowed to stand at room temperature for any length of time.

They should be kept in their case to prevent loss of moisture. Store eggs away from foods such as onions, apples, and cabbage because eggs can pick up strong odors. Store leftover raw yolks under water in a covered container in the refrigerator. Use in 2 or 3 days, or hard cook the yolks and refrigerate for up to 4 to 5 days in a tightly covered container. Leftover raw albumin or whites may be refrigerated in a tightly covered container for 7 to 10 days. Freezing is recommended for eggs that have been broken and that will not be used immediately.

QUALITY AND PURCHASING GUIDELINES

Shell color does not affect the grade, flavor, cooking performance, or nutritive value of the egg. It is determined by the breed of the hen. The purpose of the shell is to provide a safe, convenient container that is moistureproof and bacteriaproof. Buying checked or cracked eggs is a poor decision, for these eggs may be a source of salmonella bacteria, which cause food-borne infection.

When purchasing shell eggs, follow these guidelines:

1. Accept only clean, sound, and odor-free eggs.
2. Purchase eggs according to grade and size desired and only in the quantity needed for one to two weeks.
3. Accept only eggs delivered under refrigeration or cooled to temperatures below 15°C (60°F).
4. Accept only eggs packed in snug-fitting 15-dozen or 30-dozen cases of fiberboard to reduce breakage.
5. Check the grade and inspect the shells; randomly break a few.
6. Consider size and grade in relation to use and price; compare prices for different sizes of the same grade.

Size and quality are entirely different. Size of eggs refers to minimum weight per dozen. Quality refers to the grade.

The choice of the size of eggs to purchase is dependent upon the intended use. For poached or fried eggs, the smallest and least expensive egg that is consistent with quantity standards is the best choice. For omelets, batters, or scrambled eggs, the most liquid egg of good quality for the money is the best choice.

To determine the best egg buy for liquid volume, compare the prices of different sizes within the same grade. A quick way to compare value per dozen of eggs is the 7¢ rule. If the difference in price from any size to the next is 7¢, the price is the same. If there is less than a 7¢ price difference per dozen between one size and the next in the same grade, the most value is attained by buying the larger size. If there is more than a 7¢ price difference per dozen between one size and the next in the same grade, the smaller size is the better buy.

Table 8.1 gives the ounces by weight of a dozen eggs (including shells) and the ounces by liquid volume (shells not included):

COST COMPARISON

To compare the cost of different sizes of eggs, divide the cost of a dozen eggs by the liquid volume. This is the cost per liquid ounce.

To decide whether to buy small eggs for 88¢ per dozen or extra large eggs for $1.30 per dozen, determine which would be the best buy. Use the following formula to calculate the cost per liquid ounce:

$$\frac{88}{16} = 5.5\text{¢ (small)} \qquad \frac{1.30}{24} = 5.4\text{¢ (extra large)}$$

Table 8.1. Egg size, weight, and liquid volume equivalents

Egg size	Weight	Liquid volume
Jumbo	30 oz	26.7 oz
Extra large	27	24.0
Large	24	21.4
Medium	21	18.7
Small	18	16.0
Peewee	15	13.4

EGG PRODUCTS

Egg products are processed or convenience forms of eggs obtained by processing eggs. They include frozen, refrigerated liquid, dried, and specialty egg products. Processed egg products are virtually indistinguishable from fresh eggs in nutritional value, flavor, and most functional properties. By law, all egg products are processed in sanitary facilities under supervision of the USDA and bear the USDA inspection mark. They must be pasteurized and are routinely sampled and analyzed for salmonella. Products found to be contaminated with salmonella are not permitted to move in consumer channels. Egg products may be contaminated if the container seal is broken or the products are not properly handled.

The advantages of using egg products in a foodservice establishment include convenience, labor cost savings, product quality, stability, uniformity, minimal storage requirements, and ease of portion control. To avoid problems, purchase only pasteurized egg products; specify egg products bearing the USDA inspection mark; specify the exact type of egg product desired; specify the packaging desired; and do not accept frozen products that appear to have thawed.

Frozen eggs are available in a variety of sizes. Popular container sizes are 30# pails, a case of 4 10# pure-pack waxed cartons, a case of 6 5# cans, a case of 12 1-quart cartons, and a case of 24 ½-pint containers.

Frozen products should be transferred to refrigerators or freezers immediately upon delivery. When defrosting, do so quickly. Leave in the refrigerator to thaw or set in cold running water. The carton or container should remain tightly sealed. Never thaw at room temperature. Defrosting at temperatures higher than 8°C (45°F) can cause curdling and off-flavors. Use defrosted eggs promptly. Refrigerate any unused portion, and use within three days.

The following information and calculations illustrate how to compare the cost of preparing 200 2-ounce servings of scrambled eggs from small eggs, extra large eggs, and frozen egg products. Small eggs cost 88¢ per dozen, extra large eggs cost $1.30 per dozen, and frozen eggs are available in a case of 4 10# cartons at a cost of $36.00 (90¢/pound). It takes one minute to break 10 fresh eggs and one minute to open and empty a container of frozen eggs. The salary is $7.00 per hour. The cost is calculated as follows:

Small eggs: $\dfrac{200 \times 2 \text{ oz}}{16 \text{ oz}}$ = 25 doz $\times$ 88¢ = \$22.00

$\dfrac{25 \text{ doz} \times 12}{10 \times 60}$ = .5 hr $\times$ 7.00 = $\underline{\quad 3.50}$

Total cost = \$25.50

Extra large eggs: $\dfrac{200 \times 2 \text{ oz}}{24 \text{ oz}}$ = 16.67 doz $\times$ 1.30 = \$21.67

$\dfrac{16.67 \text{ doz} \times 12}{10 \times 60}$ = .33 hr $\times$ 7.00 = $\underline{\quad 2.31}$

Total cost = \$23.98

Frozen egg product: $\dfrac{200 \times 2 \text{ oz}}{16 \text{ oz}}$ = 25 lb $\times$ 90¢ = \$22.50

$\dfrac{1}{60}$ = .0167 hr $\times$ 7.00 = $\underline{\quad .12}$

Total cost = \$22.62

Frozen scrambled-egg mix is sold in a wide variety of containers ranging from a single serving to 30# pails. Low-fat milk solids are added. Some processors have other ingredients such as stabilizers and seasonings listed on the container. This product makes a high-quality, very acceptable serving of scrambled eggs.

Refrigerated liquid egg product has the advantage of convenience since there are no shells to remove and discard. Often an added incentive is reduced price. The major disadvantage is short shelf life. It must be refrigerated immediately upon delivery: always store at 5°–10°C (40°–50°F), keeping the seal intact. Liquid egg products may be kept up to 10 days unopened in the refrigerator. Once opened, use immediately. Because of the limited shelf life, liquid egg products should be purchased directly from a local packer or egg processor.

Dried eggs are used chiefly as ingredients in bakery products. They are available as whole eggs, yolks, or whites. Each of these is available in forms ranging from spray-dried to dehydrated. Dried egg mix consists of not less than 51 percent whole egg solids, not less than 30 percent nonfat milk solids, not less than 15 percent vegetable oil, not more than 1 percent salt, and not more than 3 percent moisture.

Unopened bags of dried eggs should be stored in a cool, dry place away from light. Opened bags should be resealed and stored in the refrigerator. If combined with dry ingredients and held for storage, seal the product tightly in a closed container and store at 0°–10°C (32°–50°F). After reconstitution, dried eggs should be treated as fresh eggs and used immediately in cooked products such as scrambled eggs, casseroles, or similar products.

Specialty egg products include prepeeled hard-cooked eggs, egg rolls (or "long eggs"), frozen omelets, egg patties, quiches, quiche mixes, scrambled eggs, and fried eggs.

Precooked and prepeeled No. 1, or perfect, eggs and salad eggs should be considered if labor is an important factor. No. 1 egg processing is done in such a way as to perfectly center the yolk, prevent a green ring around it, and increase the shelf life to six weeks. No. 1 eggs have no torn egg white or exposed yolk. Perfect eggs are also available in a preshaped elongated egg that supplies a large number of perfect slices for use as garnish. Salad eggs are broken pieces with at least 20 percent perfect whole eggs. For many facilities this is a good ratio of perfect eggs to broken eggs for salads and creamed dishes.

Frozen precooked scrambled eggs are available in a 1¾-ounce patty or in a folded 3½-ounce omelet. Both can be reconstituted in a microwave oven.

PASTEURIZED EGGS

Eggs can be sterilized with high-pressure injections of pure oxygen and ozone to destroy bacteria. The result is a pasteurized egg that may keep for two or three years without refrigeration. The egg has a slightly darker shell, a larger, brighter yolk, and a less runny white when cracked, and it is fresher smelling. The sterilization process is called "HyPaz" for Hyper Pasteurization.

9 DAIRY PRODUCTS AND ALTERNATIVES

IN THE UNITED STATES and some European countries, dairy products and alternatives account for over 15 percent of the food budget. Milk is a good source of protein, calcium, and riboflavin; whole milk is a good source of vitamin A, making milk a good nutritive value.

Pasteurization, a process that eliminates harmful (pathogenic) bacteria, was a requirement of the dairy industry long before grades were established or enforced. Only Grade A pasteurized milk can be sold in fluid form for consumption as a beverage or in cooking. Grade A milk must be produced according to U.S. Public Health Ordinance guidelines, which are enforced by state and local sanitarians. The "Quality Approved" shield may be used on dairy products that are of good quality and are made under USDA supervision in a clean plant.

Homogenized milk is processed to break down the fat globules into minute particles that remain suspended in solution. This process increases the shelf life and the digestibility of the product.

MILK

Whole milk usually contains 3.5 percent milk fat and 8.5 percent nonfat milk solids. Some specialty products advertising richness will contain 4.5 to 50 percent fat. Low-fat milk contains about 2 percent milk fat and has 10 percent nonfat milk solids. Skim milk has had most of the fat removed and contains less than 1 percent milk fat. It may or may not have nonfat milk solids added.

Low-fat milk is fortified with vitamins A and D, since these vitamins are lost when fat is removed from whole milk. Fluorescent lighting destroys the benefits of fortifying milk. Skim milk loses up to 70 percent of vitamin A, and 2 percent milk loses up to 35 percent after exposure. Skim milk can develop an off-flavor when exposed to fluorescent light for as little as 6 hours. After 12 hours of exposure, the flavor is unacceptable. To assure that your milk retains its vitamins and flavor, buy it in opaque containers, which block fluorescent light.

Bulk milk can usually be purchased in 3-gallon, 5-gallon, or 6-gallon containers more economically than in smaller containers. It is available in 1-gallon, ½-gallon, quart, pint, ⅓-pint, ½-pint, or 4-ounce containers, which may be paper or plastic. The various sizes should be evaluated in terms of cost and convenience.

Chocolate milk, chocolate-flavored milk, and dairy drink are made by adding chocolate flavoring materials to milk. The term "chocolate milk" is used when flavorings are added to whole milk. If the flavorings are added to any other milk product, it must be called a "drink" or "beverage." Similar products are available in which the flavoring is coffee, banana, eggnog, or strawberry.

Buttermilk is skim milk to which *Streptococcus lactus* bacterial culture has been added. This process ferments the milk to obtain a thick viscosity and slightly acid flavor. If made with whole milk, the buttermilk will contain a higher ratio of butterfat.

Lactose-reduced low-fat milk and lactose-reduced low-fat milk fortified with calcium phosphate are products designed for those who have difficulty digesting milk because they lack sufficient lactase—an enzyme that breaks down milk sugar (lactose).
Lactobacillus acidophilus culture is added to low-fat milk to produce sweet acidophilus milk for people who have difficulty digesting untreated milk. It does not have the flavor, aroma, or thickness of yogurt.

Yogurt is another milk product that is cultured. *Streptococcus thermophilous, Lactobacillus bulgaricus,* and/or *L. acidophilus* are added to milk to produce a thick, creamy product with a strong acid flavor and characteristic aroma. It may be made with whole milk, low-fat milk, or nonfat milk. It is available in frozen, frozen on a stick, semisolid, or liquid forms. Yogurt may be served plain or natural; blended with flavorings such as coffee, lemon, or vanilla; or served with nuts or fruit added during packaging. Dips can be made from yogurt by adding flavorings such as taco seasonings, dry soup mix, cheddar cheese, or bacon.

Sour cream is light cream to which a bacterial culture has been added. It has a fat content of about 18 percent. Container sizes range from 1 ounce for an individual portion and 5 ounces for retail to 5# or 10# for quantity preparation. Sour cream is available plain or can have a variety of flavorings added to it.

ALTERNATIVE MILK PRODUCTS

Dry milk, evaporated milk, and condensed milk are processed milk products. They may be used as recipe ingredients and are desirable because of their convenient storage, longer shelf life, and economical cost.

Nonfat dry (NFD) milk is pasteurized fluid milk from which fat and water have been removed; protein, vitamins, and minerals are retained. Regular NFD milk has small particles that require either reconstitution with a whip or separation by combining with other dry ingredients in a recipe. Instant NFD milk has larger particles or flakes that reconstitute easily.

Evaporated milk is unsweetened milk with about 60 percent of the water removed by evaporation. It may be diluted one part water to one part evaporated milk and substituted for whole milk. It may be used as a substitute for cream if left undiluted.

Condensed milk is evaporated milk that has been sweetened with sugar. It is used in desserts, candies, and beverages.

Filled milk (imitation milk) looks and tastes like regular milk but usually costs less. Filled milk is made from skim milk and vegetable fat instead of butterfat. It contains less nutritive value than regular milk.

CREAM

     Cream is separated from milk through the gravitation process or by using centrifuge separators. Cream is classified according to the percentage of butterfat. Varieties of cream range from half-and-half to whipping cream. Half-and-half contains 10 to 12 percent fat with added milk solids. Light cream or coffee cream contains 16 to 20 percent fat, and all-purpose cream contains 20 to 25 percent fat. Whipping cream contains 30 to 40 percent butterfat. It is usually used in foodservice for the production of whipped cream. Cream is usually packaged in pint or quart waxed cartons. Coffee cream may also be available in individual portions. Whipping cream is available in ½-pint cartons.

     Coffee whitener and whipped topping are generally nondairy or imitation cream products sometimes substituted for their real dairy counterparts. They are formulated from a combination of vegetable fats and are often less expensive and have a longer shelf life than dairy products.

ICE CREAMS, SHERBETS, AND ICES

     The decision about whether to purchase bulk desserts or individual portions will depend on a comparison of cost including labor and dishwashing, convenience, portion control, and product acceptability. The savings realized by buying bulk desserts may be negated by poor dipping techniques or deterioration resulting from lengthy storage time. Preportioned frozen desserts provide portion control.

     Ice cream is a blend of dairy products, sweetening agents, stabilizers, emulsifiers, and added flavoring materials. It has been frozen and whipped. The three categories of regular commercial ice cream are competitive, containing 10 percent fat; average, containing 12 percent fat; and premium, containing 15 percent fat. As the fat content increases, there is usually an increase in the quantity of flavoring material used. Also, as the fat content increases, the amount of overrun decreases. Overrun is the amount of air incorporated during the freezing and whipping process.

French ice cream generally has a higher butterfat content, containing more eggs and flavoring, and is a deeper yellow color when vanilla flavored. Fancy ice cream and ice cream novelties include such items as pies, cakes, spumoni, sandwiches, cups, nut rolls, molds, slices, and ice cream on a stick with a variety of centers and coatings. Most are ready to serve and individually wrapped.

Ice milk contains 3 to 8 percent butterfat, which is less than in ice cream. It has nonfat milk solids added during processing and can be flavored and made into novelties in the same way as ice cream. Ice milk is often less expensive than ice cream. Both are available in a range of container sizes. Bulk products may be packed in 3-gallon, 5-gallon, or 6-gallon containers. They may also be purchased in gallons, ½-gallons, quarts, pints, and individual cups.

Ice cream and milk shake mixes are available for facilities that have access to a soft serve machine or shake maker. Soft serve or ice cream mix has from 1 to 4 percent butterfat but can be ordered at any fat level if the order is large enough to make it economically feasible for the supplier to prepare the mix to specifications. It is usually packaged in ½-gallon or 5-gallon containers. Shake mix contains 3.5 percent butterfat and is available in 1-gallon or 5-gallon containers.

Sherbet is made from milk with fruit juices and other flavorings added. It may or may not contain egg white and is lower in fat than ice milk. It is available in bulk, gallons, ½-gallons, quarts, pints, and individual cups.

Water ice is made from water and sugar with fruit juices or other flavorings added. It contains no milk. It is usually available only as a special order and can be packed in whatever container size is specified.

BUTTER AND MARGARINES

Butter is made from sweet or sour cream and contains not less than 80 percent butterfat. A lactic acid culture may be added before churning. The addition of salt and coloring is optional. Butter that has been officially graded by the USDA bears a shield on the package with a letter that indicates the quality at the time of grading. Grades of butter relate to the quality factors of flavor, body, texture, color, and salt. Grade AA is score 93 butter, Grade A is score 92, Grade B is score 90, and Grade C is score 89. Butter is available in cuts packaged 72 to 90 per pound on parchment sheets and in prepackaged individual portions packed 60 to 72 per pound. Individual portions may be in cups, foil wrapped, or readies, which are cuts on individual cardboard squares topped with waxed paper. Four-ounce or ¼ sticks, 1# blocks, and 68# blocks are also available.

Whipped butter has been stirred and whipped to incorporate air, increase the volume, and make it easier to spread. It may be salted and may have water added. It is sometimes called "low calorie" because the volume is increased with air and water. It is available in individual portions and in 8-ounce, 12-ounce, 1#, and 5# tubs.

Margarine is made from refined vegetable oils or a combination of animal fats and vegetable oils emulsified with cultured milk, sweet milk, nonfat dry milk solids, water, or a mixture of these. The product is cooled and mechanically kneaded to produce the desired consistency. Color and butter flavoring or butter and vitamins are added. The addition of salt is optional. The law requires that, like butter, margarine must contain 80 percent fat unless it is labeled "imitation" or "diet." The label must state the fat or fats used. Regular margarine is available in individual portions and ¼# or 1# prints.

Whipped margarine, like whipped butter, is regular margarine that has been whipped. It is available in individual portions and ½#, 1#, and 5# tubs.

Soft margarine, which is less hydrogenated, is available in individual portions, ½#, 1#, and 5# tubs.

Liquid margarine, which is even less hydrogenated, is available in ½# and 13-ounce plastic squeeze bottles and in 5# plastic bottles.

10 CHEESE

CHEESE COMES IN MANY FORMS and more than 400 varieties of flavors and textures. Natural cheese is made from cow, sheep, or goat milk or cream and is usually cured or aged for a specific period to develop flavor. It is prepared by coagulating milk and separating the curd or solid portion from the whey or watery portion. Cheeses must contain a minimum of 50 percent milk fat and a maximum of 39 percent moisture. They may be classified by texture or consistency and the degree or kind of ripening. Ripening is the change over time in physical and chemical properties such as aroma, flavor, texture, and composition. Cheese can be hard, firm, semisoft, and soft. It can be ripened with mold or bacteria.

HARD CHEESES

Parmesan and Romano are the hard cheeses. Both are bacteria culture and enzyme ripened and are usually available in the grated form but can also be purchased in cuts from specialty import shops. In the grated form they are often blended together. Both are light yellow and granular and have a sharp, distinctive flavor. They are available in retail packages and 5# containers. Fully cured, they can keep almost indefinitely.

FIRM CHEESES

Cheddar and Swiss are the two most popular varieties of firm cheese in the United States. Most firm cheeses are bacteria ripened.

Cheddar cheese is the only variety that is graded. The USDA grades are based on taste, texture, and appearance. The grades most often sold are U.S. Grade AA and U.S. Grade A. Grades B and C contain possible defects. Cheddar ranges in color from pale yellow to orange and in flavor from mild to very sharp. The length of storage (ripening) time influences the flavor. The longer the curing period, the sharper, fuller, and richer the flavor. The cost generally increases with the aging period. Cheddar is available in a variety of shapes and sizes. The long cylinders, called "longhorns," are 12# to 13#. Bricks, either square or oblong, are usually 5# to 10# but are available in the larger blocks from which the bricks are cut. The bricks may be sliced. Wheels may be as heavy as 60#.

Swiss cheese is pale yellow and has characteristic holes caused by gas bubbles formed during ripening. It has a mellow, nutlike flavor. It originated in Switzerland but domestic Swiss cheese is now also available. It may be purchased in loaf, wheel, or cut form, or it may be shredded. The loaves are usually recut to 5# sizes. After cutting, the edges will dry rapidly if not securely wrapped in airtight wrap. Baby Swiss is a similar product with very small holes.

Colby is mild to mellow flavored and has an open texture. The color ranges from yellow to orange. It comes in 5# blocks or flattened rounds.

Provolone has a smooth plastic body. It is creamy white with a mild flavor that develops to sharp, piquant, or smoky when aged six or more months. It may be purchased as a tube or ball in a variety of sizes. For curing, it is paraffined or placed in plastic bags.

SEMISOFT CHEESES

Mozzarella, Brick, Muenster, Monterey Jack, Edam, and Gouda are semisoft cheeses that are bacteria ripened. Blue and Roquefort are semisoft and mold ripened.

Mozzarella is a fresh cheese with little or no ripening. It was originally made in Southern Italy from buffalo milk. The domestic variety is made from cow's milk. It has a smooth plastic body that is creamy white with a mild, delicate flavor. The most common purchase unit is a 5# loaf. String cheese is mozzarella in the shape of a rope.

Brick is pale yellow and ranges from mild to sharp. It is available in blocks the size and shape of a brick, or it may be sliced.

Muenster is a medium yellow mild and creamy cheese of German origin, available in a loaf or sliced.

Monterey Jack is smooth-textured and very creamy. Its flavor ranges from mild to mellow. It is often blended with Colby, forming a marbled appearance.

Edam is a Dutch pressed cheese that is deep yellow, made in loaf or flattened balls, and covered with red wax.

Gouda is a flattened round and is higher in fat.

Blue cheese gets its name from the color of the mold that gives the cheese its tangy, peppy flavor. It is crumbly with a sharp, piquant, spicy flavor. Imported brands are often spelled "bleu," the French word for blue.

Roquefort is an imported cheese made in France. According to the laws and regulations of France, the term "Roquefort" can apply only to this blue-mold cheese made from sheep milk in the immediate vicinity of Aveyron in southeastern France. Similar types of imported blue-mold cheese made from cow's milk are bleu from France, Gorgonzola from Italy, Stilton from England, and Danish blue from Denmark.

SOFT CHEESES

Soft cheeses include Brie and Camembert, which are mold ripened; Limburger, which is bacteria ripened; and cottage cheese, ricotta, feta, cream cheese, and Neufchatel, which are unripened.

Brie has a creamy yellow interior and is encased in a thin whitish edible crust. It has a mild to pungent flavor.

Camembert has a creamy yellow almost liquid interior in a thin grayish
edible crust. It has a piquant flavor and, like Brie, is of French origin.

Limburger originated in Belgium. It is light yellow and semisoft and has a pungent flavor. It is available in pieces or as a spread.

Cottage cheese, also called "pot cheese" and "schmierkase," has a mild, slightly acidic flavor. Cottage cheese comes in large or small curd; plain; creamed or dry; with or without salt; and with fruit, chives, onions, or vegetables added. It is available in retail packages of 5 and 8 ounces and in 5# or 10# plastic cartons or 25# pails.

Ricotta is similar to cottage cheese but is of Italian origin and is smoother.

Feta cheese is from Greece. It also is similar to cottage cheese but is drier and has a slightly salty flavor.

Cream cheese is a very pale yellow soft cheese that deteriorates rapidly. It is smooth and buttery with a mild, slightly acidic flavor. It is available in 3-ounce and 8-ounce foil-wrapped packages of retail size and 2# and 5# packages as well as jars or foil dishes of plain cheese or cheese mixed with seasonings for use as a dip.

Neufchatel is the same as cream cheese except it is lower in fat and has more moisture.

CHEESE BLENDS

Cheese blends include process cheese, cheese food, cheese spread, and cold-pack cheese.

Process cheese is not USDA graded but is often made under USDA supervision and may be marked with the "USDA Quality

Approved" shield. Process cheese is a mixture of several natural cheeses that have been shredded, melted, and pasteurized (to stop the curing at the desired stage) and mixed with an emulsifier. Flavoring materials such as pimiento, fruit, vegetables, meats, spices, or nuts may be added. It may be stored for a longer time than most cheese because it is poured into sanitary airtight plastic loaf packages, each weighing 5# or 10#. It is less expensive than most varieties of natural cheese and has the advantage of having no waste because there is no rind. Because of the added moisture and the emulsification of the fat, it melts easily and can blend into a sauce.

Cheese food or pasteurized process cheese is a product made from a mixture of one or more varieties of cheese with milk solids, salt, and up to 3 percent emulsifier. The moisture content is higher than that allowed for process cheese. The butterfat content is lower, and there is less cheese, but milk and whey solids have been added to make it softer and more spreadable.

Cheese spread is similar to cheese food except that a stabilizer is used. The moisture content is higher, and the butterfat content is lower. Spreads may be flavored with pimiento, olives, onions, pickles, or other added ingredients.

Cold-pack cheese, or club cheese, is a blend of one or more varieties of natural cheese prepared without heat and with no emulsifier. Spices or smoke flavoring may be added. It is softer than natural cheese and will spread more readily.

Cold-pack cheese food is prepared like cold-pack cheese but contains added milk and whey solids, which make it milder, softer, and more spreadable. It may include fruits, vegetables, pimiento, meat, spices, or smoke flavoring. Sugar or corn syrup may also be added.

ALTERNATIVE CHEESE PRODUCTS

Low-sodium cheese is available but may be less desirable than other cheeses because of its poor flavor, rubbery texture, and short storage time. It is available in retail packages and 5# blocks. It is suggested as an alternative to cheese when sodium intake must be reduced.

Low-fat cheese is made with skim milk. Most varieties have one-third less fat than regular cheese or about 5 grams of fat per ounce. Some brands have been judged to be a bit more rubbery than desirable, so care should be taken to look for a product that is acceptable.

Low-fat cottage cheese is also prepared with skim milk. It has 4 grams of fat per 8-ounce portion.

Imitation cheese is similar to its natural counterpart in appearance. It is made of soybeans and contains no milk solids. It is lower in cholesterol and melts readily. In many states it may be served only if notice of its use is posted on the wall or included on the menu. It is used as a low-cost alternative to cheese and is used on many pizzas and frozen casseroles.

11 FRESH FRUITS

THERE IS AN ASSORTMENT OF FRESH FRUITS available to add interest to the menu and choices to purchasing. Modern storage techniques make most fruits such as apples, oranges, lemons, limes, bananas, and avocados available year-round. Most are more plentiful, are more economical, and have better flavor during their harvest seasons. Transportation and storage costs may make them too expensive for regular use when out of season.

Other fruits, such as berries, are very perishable and cannot be stored for long periods when handled by traditional methods. However, irradiation is enhancing ripening and eliminating spoilage and mold. When consumers were allowed to select between strawberries that were irradiated and clearly marked as irradiated and berries that were handled in the traditional methods, no regular strawberries were purchased until the irradiated ones were gone.

Ripening changes are brought about by enzymes in the fruit, and they continue after the fruit is gathered. Ripening causes fruit to increase in size; tissues to soften; color to change; carbohydrates to change in kind and amount; flavor to change from acid, bitter, or astringent to mild, sweet, or bland; and typical aroma to develop. Some fruits such as bananas and avocados ripen after harvesting. Others, such as most melons, do not ripen after they are picked.

To eliminate surprises when purchasing fruit, clearly define the product desired. To establish specifications, first determine how the item will be used. For example, if lemons will be used for juice, the larger the better. If they will be used for wedges, smaller fruit is more cost efficient. This will not be true however, if the very large lemons are in short supply and are very expensive.

Arrange for deliveries to be at a convenient time. Always inspect and properly store fruit immediately after delivery. Heat and time will cause produce to become limp and to ripen and decay faster.

Apples are graded Extra Fancy, Fancy, Combination, No. 1, and Utility. About 60 percent are grown in Washington State. The rest come from California, New York, and Michigan. They are harvested in the late summer and fall but are marketed year-round. Storage of apples in controlled atmosphere facilities that provide low oxygen, low carbon dioxide, and reduced temperature and humidity helps suppliers keep domestic apples crisp and firm for many months. Red Delicious is the most popular apple available. The 12 principal varieties, which along with the Red Delicious make up 85 percent of the apple production, are Cortland, Golden Delicious, Granny Smith, Gravenstein, Jonathan, McIntosh, Newton Pippin, Northern Spy, Rome, Stayman, Winesap, and York. The variety of apple specified will depend on the use. For example, Red Delicious is good for Waldorf salad or to serve whole. Jonathans or Winesaps may be a better choice for cobbler or pie. Winesaps are especially flavorful for making cider.

Use Table 11.1 to determine which is best for each purpose.

Apples are packed in 38# to 42# cartons, if loose, and 40# and 45# cartons, if placed in trays. They are ripe when picked and should not be accepted if immature or overripe. Apples are waxed to improve appearance and retard shriveling and dehydration. The wax is easily removed when the fruit is washed before preparation. They should be ordered by count, which means the number of apples per bushel or standard carton. The count ranges from 40, which is a very large apple, to 216, which is very small. If they are to be used as an ingredient, the most cost efficient choice is the largest apple. If they are

Table 11.1. Apple varieties and best uses

Snacking	Fruit Salads	Baking
Red Delicious	Red Delicious	Rome
Golden Delicious	Golden Delicious	York
McIntosh	McIntosh	Jonathan
Granny Smith	Stayman	Stayman
Stayman	Cortland	Winesap
Winesap	Northern Spy	Northern Spy
Gravenstein	Granny Smith	Gravenstein
		Granny Smith

to be served whole and fresh, there will be more servings per case with small apples.

Apricots are graded No. 1 and No. 2. They are marketed from June through September but are most plentiful in June and July. The three principal varieties are Royal, Tilton, and Moorpak. King, Derby, Perfection, and Blenheim are also popular. Golden yellow color, plumpness, and firmness are an indication of quality and ripeness. Apricots are packed in cartons of 12#, Brentwood lugs of 24#, LA lugs of 24# to 28#, and Sanger lugs of 24# to 26#. Size is indicated by pieces of fruit in each row. There are about 12 medium apricots in a pound. Apricots are an excellent source of vitamin A.

Avocados are graded No. 1, Combination, No. 2, and No. 3. They are marketed year-round, with peak seasons from October through December and from February through April. They are available during the summer months, but the supply is low and the price high. They should be held at room temperature or in the ethylene ripening room until they yield to the touch. Allow for ripening time when ordering them green. The principal varieties sold in the United States in order of volume grown are Hass, Fuerte, Bacon, Zutano, and Reed from California and Booth and Lula from Florida. The shape of the fruit can be oval, round, or pear shaped. Skins range in color from bright green to purple to black and in texture from smooth to rough and nubby. Flesh color ranges from all green to green under the skin with beige or yellow near the seed. Avocados are packed in flats of 13# to 16# with one layer or 22# to 25# with two layers. Sizes range from 8 to 30 per layer. Very large fruit are sometimes packed in bruce boxes that are 32#. Avocados are the only fruit that is high enough in fat to be an important source of energy. They are an excellent source of potassium and a fair source of vitamins A and C.

Bananas are graded No. 1 and No. 2. They are marketed all year, since most are imported from Central America. The principal varieties are Cavendish and Valery. Bananas are picked mature and held in a controlled atmosphere until ripening is desired. At that time temperature, humidity, or ethylene gas is used to preripen the fruit. Bananas may be ordered green, to be ripened after delivery if space allows. Greentips need to be used immediately. Banana skins turn brown when they are refrigerated, but the edible portion is not adversely affected. If extra bananas need to be stored for baking at a later date, it may be better to remove the skins, mash, and freeze. Bananas are purchased according to size: large, medium, small, and petite. They are packed in 40# cartons. Most distributors will break a carton and sell 10# lots. Finger or petite bananas are packed 150 to a 40# box. Bananas contain more than 20 percent solids and are therefore comparatively high in calorie content. They are a good source of potassium.

Blackberries (dewberries) are graded No. 1, No. 2, and Select. They are marketed from May through September. They are picked ripe and so do not need further ripening and are packed in 24-quart crates with a weight of 36#. They are a good source of vitamin C. Boysenberries, loganberries, and Ollalie berries are varieties of dewberries.

Blueberries are graded No. 1 and No. 2. The principal varieties are Bluecrop, Jersey, and Weymouth. Marketing is from May to October. Technology in breeding is creating berries that have a more intense flavor and can be harvested over a longer period: Blueray, Tifblue, Bluette, Duke, and Elliott. Blueberries are packed in 12-pint trays weighing 11# or in 24-quart crates weighing 36#. They should be ripe when harvested and should not be accepted if they are immature or too ripe. They should be dry because moisture hastens molding. The silvery coating is not a sign of damage but is a protective coating or "bloom." Wild blueberries are smaller and sweeter; they are grown primarily in Maine. Blueberries are a fair source of vitamins A and C.

Cantaloupes are graded Fancy, No. 1, and Commercial. They are marketed every month of the year and are packed in ½-crate cartons with counts of 12 to 23 weighing 38# to 41#, ⅔-crate cartons with counts of 12 to 30 weighing 53# to 55#, and standard or jumbo crates with counts of 18 to 46 weighing 75# to 85#. Although melons are ordered by count, the heavier cartons will provide a higher yield per case. Cantaloupes are ripe and ready to use when they have a full slip (complete separation of stem from melon), a thick, corky netting, a background color change from green to a shade of yellow, and a fragrant aroma. They do not ripen further after harvesting and so need to be inspected; immature or overripe melons should not be accepted. They are a good source of vitamins A and C and potassium.

Cherries are graded No. 1 and Commercial. The principal dark sweet varieties are Bing, Lambert, and Van. They are large, round, and plump. Grown in the Pacific Northwest, they are excellent shippers. Royal Anne and Rainier (Golden Bing) are light sweet cherries that are large and bruise easily during shipping. They are grown in the East and Midwest. Tart red varieties (pie cherries) are seldom eaten fresh. The most frequently found varieties are Early Richmond, Montmorency, and English Morello. The marketing season is from May to July. Cherries are picked ripe and do not ripen further; they just decay. Cherries are packed in 12# flats or 20# lugs and cartons with varying counts. A 13-row lug is the smallest and 9-row lug is the largest. There are about 80 medium cherries per pound. Cherries are a fair source of vitamin C.

Cranberries sold fresh are all graded No. 1. All other grades go to the processor where they are canned as jelly or sauce. The principal varieties are Early Black, Howes, Searles, and McFarlin, and they vary from small and dark red to large and bright red. Cranberries are grown in Massachusetts, Wisconsin, Washington, and New Jersey. The crop is harvested in the fall, and during fall and winter cranberries are available raw. They do not ripen after harvest. Fresh cranberries are packed in cartons of 24 1# bags or in 25# cartons. They are a fair source of vitamin C.

Grapefruit are graded Fancy, No. 1, Combination, No. 2, and No. 3. The two principal white varieties are Duncan, with seeds, and Marsh, which is seedless. The two pink varieties with seeds are Foster and Thompson. The three pink seedless varieties are Ruby Red, Marsh Pink, and Star Ruby. The seedless varieties are more expensive than the seeded ones, and the pink or ruby may cost more than the white. Good, juicy fruit will be heavy, firm, and smooth-textured. It may range in skin color from pale yellow to russet or bronze. Color does not indicate ripeness. Occasionally grapefruit undergo a process called "regreening," during which ripe picked fruit begins to turn green, but this does not indicate unripe fruit. Grapefruit are marketed all year.

They must be ripe when picked, so do not accept fruit that appears to be immature or too ripe. They are packed from September to June in Florida in 4/5-bushel cartons weighing 45# with counts of 23 to 48 or in 38# to 42# cartons with counts of 23 to 64. Texas packs from September to June in 38# to 42# 7/10-bushel cartons with counts of 18 to 56 or 1 2/5-bushel carton weighing 76# to 84# with counts of 46 to 112. Both California and Arizona pack from November to October in 38# to 42# cartons with counts of 23 to 64. Look for firm, heavy fruit with thick skins. Grapefruit are an excellent source of vitamin C.

Grapes are graded Fancy, Extra No. 1, and No. 1 and are divided into five groups: table, raisin, wine, juice, and canning. They are grown primarily in California. Table grapes are those intended for use as fresh fruit. Most customers prefer seedless grapes, but many seeded grapes have good color and flavor. Characteristic color is a good sign of ripeness. Ripe grapes will also be fairly soft and tender. They do not ripen after they are picked. Do not accept a product that is not usable as it is received. The silvery coating is not a sign of damage; it is a protective coating or "bloom." The principal varieties of white seedless are Perlettes and Thompson. The best-known red seedless are Flame and Ruby Red. The main varieties of red seeded are Tokay, Cardinals, Ribers, Red Malaga, Queen, Delaware, and Concord. Popular white seeded varieties are Calmiria, Niagara, Lady Finger, and White Malaga. Grapes are marketed every month except May. They are packed in 17# to 23# lugs, flats, or cartons with varying counts. Many varieties of grapes are relatively high in calories because they are a more-concentrated source of carbohydrates, but they are not an important source of other nutrients.

The principal varieties of grapes used for raisins are Natural Seedless, Golden Seedless, and Seeded Muscats. Raisins are graded A, B, and C. They are marketed year-round. Raisins are packed in 10#, 25#, 30#, and 40# containers with or without polybag liners.

Kiwi fruit are available year-round, coming from California in the fall and New Zealand in the spring. The principal variety is Hayward. Ripe fruit should be firm but give slightly to the touch. There should be no cuts, bruises, or soft spots. Kiwi fruit can be ripened at room temperature. Refrigerate after ripening. They are packed in 7 7/10 # cartons with 25, 28, or 30 count or 6 6/10 # cartons with counts of 33, 36, 39, or 44.

Lemons are graded No. 1, Combination, and No. 2. They are grown chiefly in California and Florida and are marketed year-round. They are mature and fully developed when picked. Do not accept fruit that is soft and overripe. They may be waxed to reduce shriveling. Lemons are packed in 37# to 40# cartons with counts of 63 to 235. The best buy is the size that is the most plentiful. They are an excellent source of vitamin C.

Limes are graded No. 1, Combination, and No. 2. There are two types of limes: sweet and acid. Sweet limes have a yellow or green-yellow skin that is smooth and elastic. Acid limes have a dark green skin. Only acid limes are grown in the United States. Acid limes are further divided into two types: Tahiti (large fruits) and Mexican (small fruits with thin rinds). Varieties of Tahiti types include Persian, Bearss, Idemore, and Pond. The Persian variety is most widely grown. They are marketed year-round but are most plentiful August to February. Packing is done in 10# flats or cartons with counts of 36 to 63, in 20# cartons or flats with counts of 72 to 126, or in 38# cartons or bruce boxes with counts of 110 to 250. They are an excellent source of vitamin C.

Melons other than cantaloupes and watermelons are graded No. 1, Commercial, and No. 2. The principal varieties are Honeydew, Honeyball, Casaba, Spanish, Crenshaw, Persian, and Santa Claus.

Honeydew melons are greenish white to light yellow. They have a velvety skin and a light green meaty flesh. Orange-fleshed honeydew are relatively new and smaller than the green-fleshed variety. When ripe, the rind is pale yellow-orange and highly perfumed. The orange flesh is crisp like cantaloupe, sweet, and succulent.

Casaba melons are shaped like giant figs and have thick, hard yellow-green skins with soft creamy white flesh. They average 4# to 6#. Ripeness is indicated by a slight softness at the blossom end. There is no odor or aroma.

Spanish melons look like a cross between a Casaba and a Crenshaw. Their hard skins are green to deep yellow, and they have deep orange flesh.

Crenshaw melons have golden netless rinds with longitudinal wrinkles. They are pointed at both ends and have a gold pink to deep orange flesh. Ripe melons will be slightly soft at the blossom end and have a pleasant aroma.

Persian melons have a green rind with a fine brown netting and a pink-orange flesh. They are similar to a honeydew but a little smaller. Persian melons are slightly soft and fragrant.

Santa Claus melons look like small watermelons with spotted green and yellow rinds. Their soft green flesh looks and tastes like the honeydew's. They are marketed from May to January. They do not ripen after harvesting, so do not accept immature or overripe fruit. Packing is done in cartons weighing 26# to 27# with counts of 6, 7, or 8 and in 29# to 30# cartons with counts of 4, 5, or 6. A new melon variety, about the size of an egg or plum, is the Eat-all melon, which has an edible rind and is seedless. So far, the one species developed has a bright yellow rind, pink flesh, and sweet melon flavor. The Eat-all is being bred for longer shelf life so that it can be marketed in vending machines.

Nectarines are graded Extra No. 1 and No. 1. There are numerous varieties including Fantasia, Flamekist, May Grand, Sun Grand, Autumn Grand, Flavortop, and Firebrite. They ripen at various times, making them available from late May to September. They are picked when mature and ripe, but time must be allowed for additional ripening to soften and enhance flavor and appearance. Nectarines are packed in 19# to 23# lugs or cartons with counts of 32 to 96, 25# lugs or cartons that are volume fill, and 35# cartons with varying counts. They are a good source of vitamin A and a fair source of vitamin C.

Oranges are graded Fancy, Premium, Choice, No. 1, Combination, No. 2, and No. 3. The principal varieties for juice are Valencia, Hamlin, and Pineapple. Pineapple oranges are usually selected for juice because they are thin-skinned. Navel are usually chosen for hand eating because they are sweet and thicker skinned, making them easy to peel. Oranges are produced in California, Texas, Florida, and Arizona. There is year-round marketing. They are mature and ripe when picked. Sometimes oranges undergo a process called "re-greening" (see the section on grapefruit). These oranges may have artificial color added to make the skin orange again. They must be marked "color added," but this does not affect the quality or safety. Oranges are packed in ⅘-bushel cartons weighing 37# to 45# with counts of 64 to 125 when shipped from California and Florida. Navels are shipped in the same size carton with counts of 32 to 56 from Florida. Texas packs oranges in 38# to 44# cartons with counts of 56 to 144 and 82# to 87# cartons with counts of 100 to 288. The best buy is the size that is most plentiful and the fruit that is heaviest for its size. Oranges are an excellent source of vitamin C.

Peaches are graded Extra No. 1 and No. 1. They are produced mainly in California and Georgia. The early type is clingstone in which the flesh is firmly attached to the pit; the later type is freestone in which the flesh separates easily from the pit. The principal varieties are Elberta and Red Haven. Others are Suncrest, Fairtime, Red Top, Springcrest, and O'Henry. New varieties are being developed each year. The skin color ranges from white to deep yellow to bright red blush. The flesh ranges from white to deep yellow. The marketing season is from May to November. The fruit is packed in 30# cartons or lugs with counts of 32 to 96, 17# to 18# boxes with counts of 35 to 70, and 30# and 38# cartons with varying counts. Peaches do not ripen after picking, so they must be mature at the time of purchase. A few days should be allowed between purchase and use for the fruit to become softer and more appealing before serving, if necessary. They are a fair source of both vitamin A and C.

Pears are graded No. 1, Combination, and No. 2. They are produced mainly in California, Oregon, and Washington. The principal varieties are Bartlett, Anjou, Bosc, and Comice. Bartletts, the best-selling variety, have bell shapes with a greenish yellow skin that is sometimes red blushed. The less green its skin, the riper the fruit. Anjous are squat-necked with yellowish green skin and usually have russeting. Boscs are yellow skinned and heavily russeted with a cinnamon color. They have long tapering necks. Each variety has a season of about three months, and because of the overlapping seasons pears are marketed every month except June and July. Pears are picked unripe since they do not develop properly on the tree. They are packed in 22# to 24# lugs or cartons and 36# cartons with varying counts or in 44# to 46# standard cartons with counts of 80 to 188. They are picked and packed green and have a better texture and flavor when they ripen off the tree.

Pineapples are graded Fancy, No. 1, and No. 2. They are grown in Hawaii, Mexico, and Honduras. The principal variety is Smooth Cayenne. They are marketed year-round with the peak period between April and June. They are picked ripe and will not continue to ripen. Ripe fruit has an intense fragrance at room temperature. It should be firm yet yield slightly to pressure and should be heavy for its size. Pineapple is packed in 20# cartons with counts of 4 to 7 or in 40# cartons with counts of 8 to 18. They are a good source of vitamin C.

Plums are graded No. 1. Most of them are grown in California, Oregon, and Washington. The two main types are Japanese, which are juicy and a variety of shapes and colors, but never blue or purple, and European, which are smaller, oval, and always blue or purple and sold fresh or dried. There are more than 2,000 varieties. Some of the more popular are Simka, Santa Rosa, Laroda, Cassleman, Red Beaut, Friar, and El Dorado. They are sized extra large, large, medium large, medium small, and tiny. The marketing season is from May to December. The silvery glaze is not a sign of damage but is a protective coating or "bloom." Plums are packed in 28# cartons or lugs with counts of 144 to 225, 18# lugs or cartons with counts of 60 to 118, or 22# to 38# crates with counts of 132 to 280. They are a fair source of thiamine.

Prunes are not graded. They are a type of plum. The main variety is Italian, grown in the Pacific Northwest. Plums are available fresh in late summer, packed in 28# ring-faced baskets. They are available dried year-round. Prunes are packed in 25# and 30# boxes either pitted or unpitted. They are valued for their laxative property and vitamin A and iron content.

Raspberries, both red and black, are graded No. 1. They are shipped from California from April to November, with peak production in June and July. Picked ripe, raspberries should not be accepted if immature or too ripe. They should be dry because moisture hastens molding. They are packed in ½-pint containers with 12 containers per flat and 3 flats per bundle. Red raspberries are a fair source of vitamin A, and both red and black are fair sources of vitamin C.

Strawberries are graded No. 1, Combination, and No. 2. Florida, Mexico, and New Jersey supply berries November to January.

California berries are available from January to November. Although they are marketed all year, peak production is from April to July. Strawberries do not ripen further after harvesting and deteriorate rapidly. They should be picked and stored with the green tops intact and should be dry because moisture hastens molding. They are packed in 12-pint trays weighing 12# with varying counts. Strawberries are a good source of vitamin C and a fair source of potassium.

Tangerines are divided into four categories: Mandarins have a mild, sweet flavor and a lighter orange color than other varieties, including Satsumas, Kinnow, Wilking, and Kara. Tangerines are small rich-flavored deep orange fruit, including Algerian, Dancy, Kinnow, and Honey. Temple oranges are larger with a pleasant tart-sweet flavor. Tangelos are a grapefruit-tangerine cross. They combine the best of both fruits for a tangy, sweet flavor. The principal varieties of tangelos are Orlando and Minneola. Each variety has a peak season of two to four months. Some combination of these varieties is available from October through May. They are fully ripe when harvested and do not ripen further. Florida packs them in ⅘-bushel crates or cartons, with counts of 176 and 210 being the most popular. California packs by size from the categories small to super colossal. They are an excellent source of vitamin C.

Tomatoes are fruit botanically, but the Supreme Court decreed in 1893 that they are vegetables, at least for purposes of trade. (See Ch. 12 for information on tomatoes.)

Watermelons are graded No. 1, Commercial, and No. 2. They are grown primarily in California and the southern states. They may vary from cylindrical to oval to round. The principal varieties are Charleston Grey, Jubilee, Peacock, and Black Diamond. A new variety is seedless, making it have a longer shelf life because watermelon seeds give off ethylene gas, which speeds ripening. This variety is round, thin-skinned, deep crimson, and sweet. Watermelons are marketed from mid-April to December. They should be harvested ripe and will not ripen during holding. Watermelons may be purchased individually or in 70# to 87# cartons with a count of 3, 4, or 5. They are a fair source of vitamin A.

Table 11.2 lists less well known fruits that might be used in some foodservice operations.

Table 11.2. Less well known fruits used in foodservice operations

Apple-pear	Gooseberry, green	Papaw
Arracacha	Granadilla	Papaya
Asian pear	Guava	Passion fruit
Atemoya	Jackfruit	Pepino
Babaco	Jonagold apple	Persimmon
Blood orange	Jujube	Physalis
Boniato	Kiwano	Pitaya
Breadfruit	Kumquat	Plantain
Cape gooseberry	Lady apple	Plumcot
Carambola	Lemongrass	Pomegranate
Cherimoya	Longan	Prickly pear
Coconut	Loquat	Pummelo
Crabapple	Lychee	Quince
Currant, black	Mamey sapote	Red Banana
Currant, red	Mango	Rhubarb
Dates	Mangosteen	Sapote
Elderberry	Monestera	Star apple
Feijoas	Oca	Tamarillo
Figs	Oriental pear	Tamarind
French butter pear		Ugli fruit

12 FRESH VEGETABLES

THERE IS A MULTITUDE OF FRESH VEGETABLES available to add calories, texture, color, and variety to the menu. This same wide array of choices adds challenges to purchasing produce. Modern storage and transportation make many vegetables available all year to most of the country. Some more-seasonal vegetables are being grown under controlled conditions. Hydroponics is a greenhouse farming technique that substitutes liquid nutrients for soil and allows produce to be grown closer to the places it will be consumed. The crops grow twice as quickly in this controlled environment as their field-grown counterparts. Through hydroponics fresher produce is available year-round and at a lower cost.

Vegetables are divided into categories according to the part of the plant. Tubers, seeds, and roots have been adapted by the plant as storage portions; they are high in carbohydrates and are important sources of energy in the diet. The actively growing portions (shoots, leaves, flowers, and fruits) are high in water content and are primarily useful as sources of minerals and vitamins. Green or succulent vegetables are highest in quality and nutritive value immediately after they are gathered.

The maturing of vegetables, like the ripening of fruit, is brought about by enzymes whose action continues after the vegetable is gathered. Vegetables dry out easily and wilt quickly after they have been picked. This is especially true of tender ones with high water content and without a protective peel or skin. Enzymatic changes and water loss are most rapid in leaves and growing portions. The most serious effect is that enzymes reduce vitamin content. The roots, tubers, and bulbs are the most stable.

Adaptability of the vegetable to storage, available storage space, nutritional needs, and the number to be served determine the quantity to be purchased.

Artichokes are graded No. 1 and No. 2 and are available every month except September. Winter artichokes have a plump globular shape with compact, tightly closed leaves and a bright green color. The main variety is Globe. Summer artichokes are naturally cone-shaped and loose and have spiky, spreading leaves. Jerusalem and Chinese artichokes are tubers and very different from the unopened flower bud of the thistlelike plant of the commonly known artichoke. Avoid artichokes with mold, discoloration, or worm injury. Bronzed outer leaves indicate they have been frost touched and will be tender and succulent. Open spreading leaves indicate overmaturity and toughness. Brown leaves indicate aging and damage. Artichokes are packed in 20# to 25# cartons with counts of 18 to 60. Small loose counts are also available. They are a good source of calcium.

Asparagus is graded No. 1 and No. 2 and is marketed between February and July, with the most plentiful supply between April and June. The two general types are green and white. Look for closed, compact tips, smooth round spears, and a fresh appearance. Stalks should be tender and green. Avoid open, seedy buds, which indicate overmaturity, moldy or decayed tips, or ribbed spears. Freshness is best maintained by storing spears standing upright in water. Asparagus is packed in 15# crates, 24# to 25# cartons, 32# pyramid crates, and 32# loose packs. Asparagus is a good source of vitamin C and a fair source of vitamin A.

Beans may be either bush or pole (vining) and are graded Fancy, No. 1, Combination, and No. 2. They may be either green (snap) or waxy yellow (wax beans). The principal producing areas are Florida, Georgia, California, Jew Jersey, New York, North Carolina, and Virginia. Beans have pods that are either round or flat. Look for crisp, straight pods with a clear color. Uniform size beans are of higher quality. Fresh beans will snap cleanly when bent. Avoid wilted, decayed, or flabby pods. Russeting in moderate amounts will not affect quality or cooked appearance. Beans are marketed year-round in 26# to 31# hampers with varying counts. Green beans are a fair source of vitamin A.

Kidney beans are the mature seed of green beans. Lima beans are flat and kidney shaped. The smaller ones are called "butter limas" and the larger beans are called "potato limas." They are available year-round, the peak period being from June through September. Beans should have a fresh, bright appearance with good color for the variety. They are packed in 10# and 20# cartons. Garbanzo beans are also known as "chick peas." Garbanzo beans are a good source of incomplete protein.

Beets are graded No. 1 and No. 2. They are shipped mainly from California, New Jersey, Ohio, and Texas year-round, with the peak season being from June to October. If they are not stored, they may be shipped with tops on, or young tender tops may be cut, bunched, and sold as salad or cooking greens. Accept those that are a rich deep red color and firm, round, and smooth with unblemished skin. A thin tap root at the bottom of the beet will usually indicate it is not woody. Avoid those that are wilted and flabby. There is a newer golden variety that is similar in texture and flavor to the traditional red beet. Beets are packed in bags of either 25# or 50# capacity. They are also available in 43# to 47# cartons with a count of 24 bunches (3 to 5 beets per bunch). Beets are a fair source of vitamin C and iron. Beet greens are an excellent source of vitamin A, a good source of both vitamin C and iron, and very low in calories.

Bok choy is a variation of the cabbage family (see cabbage).

Broccoflower is a cross between cauliflower and broccoli. It has the shape of cauliflower and the color of broccoli.

Broccoli is graded Fancy, No. 1, and No. 2. Calabrese is the most common and chief commercial variety of broccoli. Other varieties include Green Duke, Emperor, 458, 501, Shogun, and Premium. Broccoli is a member of the cabbage family. It is marketed year-round, the peak season being from October to April. Accept firm, compact clusters of buds of dark green or sage green with some purple florets. Yellow buds and stem woodiness develop as the broccoli ages. It should be used soon or stored on ice to maintain proper humidity. Broccoli is available in a wide range of packs and cuts including florets and bunches. Florets are a 100 percent usable product, with a guaranteed weight per box that doesn't fluctuate with growing conditions. The use of florets eliminates labor during preparation and cuts freight and cold storage costs by more than half. Bunched broccoli is packed in 20# to 23# cartons with counts of 14 to 18 bunches. It is an excellent source of vitamins A and C and a fair source of riboflavin, calcium, and iron.

Brussels sprouts are graded No. 1 and No. 2. They are grown mainly in California and are marketed from August to May, with peak supplies from October through December. Good sprouts are firm, green, and compact and have moist, fresh butts. They look like little cabbages. A wilted or puffy appearance is an indication of poor quality. Loose, soft, or yellow leaves indicate aging. They are packed in 12-pint trays weighing 9# per carton and in 25# cartons with varying counts. Uniformity of size is desirable. Brussels sprouts are an excellent source of vitamin C and a fair source of iron and vitamin B.

Cabbage is graded No. 1 and Commercial. Of the 300 varieties, the principal ones are Domestic, Danish, Pointed, Red, and Savoy. Domestic is the most popular variety of cabbage and has a round, compact head. Danish is tight and smooth-leaved. Pointed cabbage is early with heads of a conical shape. Red is identical to the other varieties except for its reddish purple color. Savoy has yellowish crimped leaves that form a loose head similar to Iceberg lettuce. Look for firm, heavy, fresh heads consistent with the variety. Avoid wilted, yellow, blemished, dried, or decayed outer leaves separating from the head or pulling away from the stem. Early and new cabbage is not as firm as fall and winter strains. Cabbage is marketed year-round. It is packed in 40#, 50#, and 60# cartons and bags with varying counts. Some suppliers will sell broken crates by the pound. Cabbage is a good source of vitamin C, potassium, and fiber. It is also available shredded, alone, or in combination with other vegetables in 5# and 10# cello bags.

A novel type is celery cabbage or Chinese cabbage, sometimes called "bok choy" (several varieties), which, with its long oval head, has characteristics of both romaine and cabbage. Other members of the cabbage family are brussels sprouts, collards, cauliflower, kale, and kohlrabi.

Carrots are graded Extra No. 1, No. 1, Jumbo, and No. 2. They are marketed year-round, with the main sources in California, Texas, and Michigan. They are sorted by size and shape: large, medium, or small, short tapered with a round base, or short tapered with a conical base. They should be firm, crisp, well formed, smooth, and orange to reddish orange. Wilted, flabby, shriveled, or sunburned ones should not be accepted. Baby carrots can either be immature carrots of standard varieties or mature carrots of the miniature Belgium variety. Carrots are packed in cartons containing 48 1# bags, 24 2# bags, 18 3# bags, and bags weighing 25# and 50#. Precut carrots are available shredded, chopped, diced, and in straight or crinkle-cut sticks, "coins," chunks, or pieces. All precut carrots represent a 100 percent usable product that reduces labor, freight, and storage costs. Carrots are an excellent source of carotene, a precursor of vitamin A.

Cauliflower is graded No. 1. It is a member of the cabbage family and is marketed year-round, with peak supply September through November. The size of the head does not affect quality. It should be white to creamy white, firm, and compact with fresh, crisp bright green leaves. Do not accept cauliflower that has insect injury, mold, dark spots, a ricy appearance, or a yellow-green tint. Whole heads are packed in 16# to 23# cartons, cello wrapped, with counts of 9, 12, or 16. Cauliflower florets are also available in several sizes and packs, making them laborsaving and reducing freight and storage costs. Cauliflower is a good source of vitamin C.

Celery is graded Extra, No. 1, and No. 2. Pascal is the green celery that represents 85 percent of the total grown. Celery is marketed year-round, with principal supplies coming from California, Florida, Michigan, and New York. The stalk should be straight with rigid ribs, light green, and glossy and have fresh leaves. It should snap cleanly when bent and should be clean and unridged on the inside surface. Rough or puffy stalks may be pithy. The Golden Heart variety is a bleached white and has a milder flavor. The heads are covered with soil during growing to prevent the sun from developing the natural chlorophyll coloration. Celery is packed in 25# to 28# cartons with counts of 12, 18, or 24 and in 60# crates with counts of 18 to 48. Some suppliers sell broken cases by the head. Celery is available diced, in sticks, crescent cut, chopped, or in chunks. For garnishing, it may be available by request with leaves intact on the center stalks or hearts. Celery is a fair source of vitamin C and is very low in calories.

Collards are graded No. 1. Closely related to kale, they are available year-round from Georgia and Virginia, with the best supply from January through April. They should be fresh, crisp, clean, and free from insect injury. Wilting and yellowing indicates aging and damage. They are packed in 20# to 23# cartons. Collards are an excellent source of vitamins A and C.

Corn, another name for maize, is graded Fancy, No. 1, and No. 2. Sweet corn is grouped by kernel color: yellow, bicolor, and white. Over 200 varieties are grown in the United States with most coming from Florida. Good quality corn has a fresh tight green husk with tender, milky, full kernels. The silk should be soft, moist, and free of decay or insect damage. There should be no rows of missing kernels, worm damage, or signs of decay. Large deep yellow kernels are overripe and tough. Corn that has been stored too long may have dented kernels. Soft, small kernels and wide-spaced rows indicate immaturity and lack of sweetness and flavor. New varieties have been bred to yield sweeter, longer-lasting ears and are insect resistant. Corn should be used immediately, for quality, sweetness, and flavor deteriorate rapidly. Packing is done in 45# to 50# crates with counts of 54 to 66. Corn is a fair source of vitamin C and thiamine.

Cucumbers are graded Fancy, Extra No. 1, No. 1 Small, No. 1 Large, and No. 2. They are marketed year-round from Florida, North Carolina, South Carolina, and California. There are numerous species of cucumbers of various colors, sizes, skin textures, and times of maturity. They should be firm, crisp, and of characteristic color, not soft and flabby. They may have hidden damage caused by mishandling or poor growing conditions, such as freezing temperatures or excess rain. Older cucumbers are dull green to yellow and overly large and may be woody. Cucumbers are washed, graded, and then waxed to retard evaporation before shipping. Genetically altered cucumber plants have genes designed to protect them from disease. The European or English cucumber is an almost seedless variety that is shipped shrink-wrapped in plastic. Cucumbers are packed in 26# to

28# cartons, 30# to 32# cartons, and 50# to 55# bushel cartons. They are available sliced in 5# cello bags. Cucumbers are a fair source of vitamin C.

Eggplants are graded Fancy, No. 1, and No. 2. They are grown in Florida, New Jersey, and California. Most are a deep purple color, but other varieties with other colors are available. Purple eggplants should be clear, dark, glossy, and heavy for their size and have firm flesh. Shape can vary from oval to globular. Choose small to medium size eggplants. Avoid those that are soft, shriveled, cut, poorly colored, or marked by brown spots. Eggplants should be used immediately and not stored. They are packed in 20# to 22# lugs with 24 count and in 30# to 34# bushel cartons with 30 and 35 count.

Endive, Belgian, is graded No. 1 and imported from Belgium. Also known as "witloof," it is in the same family as the more familiar curly endive. Its white color is the result of being grown without light, which prevents the development of chlorophyll. Endive is available from October to April. It should have a tight head and be creamy white blending to yellow at the leaf tips. The bitter cone-shaped base should be removed. Avoid limp, wrinkled leaves and loose heads. It is shipped and stored in blue paper packaging to prevent light from causing it to turn green and lose its delicate flavor.

Endive, curly, or chicory, is graded No. 1. Curly endive grows in a bunchy head with narrow ragged-edged leaves that curl at the ends. The outer leaves are bitter; the inner ones are sharp and tangy. It should not be confused with escarole, which has broad leaves. It is marketed year-round. Look for fresh, crisp, tender, semiglossy leaves with good green color at the edges to almost white at the center near the stem. Do not accept heads with brown or yellow discoloration. The Italian red chicory is more commonly known as "radicchio." Packing is done in 35# to 40# cartons with 24 count. Endive wilts quickly at room temperature. Buy and use it immediately. It is a good source of vitamin A.

Escarole is a broad-leaved variety of Batavian endive but is less curly at the tips. The heads are looser and the leaves less ruffled. Not as bitter in flavor as endive, escarole works well in salads. It is marketed year-round. The leaves should be bright green, crisp, fresh, and tender. Yellow or brown leaves are a sign of age and should not be accepted. The packing is done in 35# to 40# cartons with 24 count. It is a good source of vitamin A.

Greens include turnip greens, collards, mustard greens, spinach, dandelion greens, kale, beet greens, chard, and Swiss chard. Select crisp, fresh-looking greens that are free from defects. Store in a cool, moist place and use as soon as possible. They are available in units ranging from a single bunch to cases of various sizes, depending on the supplier, the product, and the season. Storage with crushed ice will prolong freshness. Greens are an excellent source of vitamins A and C.

Kale (a variation of the cabbage family) is graded No. 1 and Commercial. It has a large, hardy, curly-leaved loose head and is grown extensively in Virginia, Maryland, and New York. Scotch kale has yellow-green leaves that are curly and ruffled across the entire surface. Blue kale has blue-green leaves that are curly at the edges but are flatter and smoother in the middle. It is marketed year-round but is most abundant in the winter. Leaves should be crisp, clean, and free from bruises. A slight edge yellowing is natural and will not affect the quality. During the winter months, kale occasionally develops a bronzed appearance from frost. This does not affect the quality. Kale is packed in 18# to 25# cartons with varying counts.

Lettuce is graded Fancy, No. 1, and No. 2. The general types are crisphead (Iceberg), butterhead (Boston and Bibb), cos (Romaine), leaf (bunching), and stem.

Iceberg, the most popular of all salad greens, is grown almost exclusively in California and Arizona. Heads should be firm but not hard, free from burned or russeted tips, and fresh. The butt should be milky white to light brown. The outer ribs should not show any pink color. Russet spots generally indicate overmaturity. Lettuce will brown if stored with foods that emit ethylene gas such as tomatoes, apples, citrus, bananas, and melons. It is marketed year-round. Packing is done in 44# to 52# cartons with counts of 18 to 30 and in 38# to 45# cartons with counts of 18 to 38. It is shipped under refrigeration, which retards deterioration but accelerates dehydration; to avoid dehydration, it may be shipped vacuum cooled.

Cleaned and trimmed lettuce is stripped of its outer leaves and washed before shipping; it has a long shelf life. It is packed six heads to a bag, four or five bags per box. Some brands are vacuumed, flushed with a gas atmosphere to remove bacteria, and then heat sealed.

Cored and trimmed lettuce also has the butt removed, which shortens its shelf life but reduces the preparation time.

Shredded Iceberg is available in 5# and 10# bags for salads, with or without a mixture of other shredded vegetables.

A new small head the size of a tennis ball has been developed that either is green or has blush red outer leaves.

Butterhead varieties have soft pliable leaves and a delicate buttery flavor. Principal varieties are Big Boston, White Boston, Bibb, and May King. Bibb (Limestone) lettuce is smaller and darker green than Boston. It has small rosettes of green and yellow leaves that are soft and velvety with a delicate buttery flavor and sweet and tender with a chewy texture. When mature its leaves are short, spatulate, and closely clustered.

Cos has a loaf-shaped head and long leaves. The leaves appear coarse but are tender; the taste is strong but not bitter.

Boston lettuce is often mistaken for Bibb because it has a similar round soft head, but it is larger and not as delicate. It has leaves that are loosely packed and easily separated. The heads are fragile and very perishable.

Leaf lettuce grows with flat or curly leaves loosely branching from its stalk. It is available in a number of varieties and colors ranging from green to reddish bronze. It has a crisp texture but is tender and fragile with a delicate and sweet flavor. The enlarged stem or seed stalk of stem lettuce may be peeled and eaten raw or cooked. Celtuce is the only variety sold in the United States. The very green leaves of lettuce are a fair source of vitamin A.

Mushrooms are graded No. 1 and No. 2. The color ranges from white to off-white and cream to light brown. The caps should be dry and firmly attached to the stem. The most popular variety is the Agaricus, or Button, because of its availability and mild flavor. Mushrooms are marketed year-round, with peak supply in November and December. They should be handled gently, for aging and bruises cause darkening. Deterioration is accelerated with plastic wrap or airtight plastic bags. Withered mushrooms should not be accepted. Most suppliers provide three sizes of the product to suit your needs: small, 89 ± 12 per pound; medium, 43 ± 7 per pound; and large, 23 ± 3 per pound. They are packed in 3#, 5#, and 10# cartons.

There is a wide variety of exotic mushrooms, varying in shape, texture, and flavor. The enoki, or golden mushroom, is a long-stemmed white mushroom with a small cap and crisp texture. The taste is unusual and fruity, ideal for salads and stir-fry. The oyster is a fan-shaped mushroom with a wonderful texture and flavor. The deep-gilled caps narrow at the base to short stems. They can be used raw or cooked. The shiitake, or oak mushroom, is large, dark, and open capped with a full, lusty flavor. It is very common and often substituted for the button mushroom. The crimini, or Italian Brown, looks like a large button mushroom but is much larger and brown. It is more flavorful than the others and is best served cooked. Trumpet-shaped chanterelles provide another distinctive flavor. They have a rich yellow color and pleasant aroma. Cepes and porcini have a woodsy distinctive flavor and are usually cooked. Morels are spring mushrooms that cannot be cultivated and therefore are very expensive.

Mustard greens are graded No. 1. They are available year-round and are grown mainly in the South. The leaves should be fresh, tender, crisp, and green. Wilted, dirty, discolored, or spotted greens should not be accepted. The young tender leaves can be used in salads; the older tender leaves are used for cooking. The greens are packed in 18# to 25# bushel cartons with varying counts.

Okra is graded No. 1. Good quality okra may range in color from deep to light green. The pods may be long and thin or short and chunky. They may be smooth or ridged. They should be young, tender, fresh, and clean. A dry, dull appearance results after extended storage. Do not use an iron, copper, or tin utensil to cook okra, or it will discolor. Okra is marketed year-round and is shipped in 10# and 20# baskets. Okra is a good source of vitamin C.

Onions are grouped into two main types, dry and green. The dry category includes three varieties: the mild or all-purpose onion, which includes the Bermuda, Granex, Grano, and Spanish, is graded No. 1,

Combination, and No. 2; the pungent cooking onion, or the Globe family, is graded No. 1, Combination, and No. 2; and the strong-flavored type such as the Fiesta and El Capitan is graded No. 1, Export, Commercial, No. 1 Boilers, No. 1 Picklers, and No. 2. Sweet onions are very similar in size and shape to other dry onions but are grown in different areas and available at different times. The best known are the Maui, Vidalia, and Walla Walla, named for the areas where they are grown. Cocktail onions are grown crowded together and harvested early to keep them from becoming too large. Dry onions are marketed year-round. Good quality onions are firm with small tight necks and papery outer scales and are free from sunburn spots. Do not accept those with thick, hollow, woody centers in the neck or with fresh sprouts. They are packed in 10#, 25#, and 50# bags and in 48# to 50# fiberboard cartons with varying counts. Specify medium or large size. Dry onions are also available minced, diced, chopped, or in chunks packaged in 5# cello bags.

Green onions, or scallions, are graded No. 1 and No. 2. They have a definite bulb formation that is white about 3 inches above the root and have bright green tops. True scallions usually have a milder flavor and finer texture than the standard onion. Green onions are packed 48 bunches to a 25# carton. All varieties are marketed the year around.

Shallots, also sometimes called "scallions," are a cluster of immature, undeveloped cloves stemming from a single root. They are also available dry. Leeks have flat leaves and a white stalk about 1½ inches in diameter and 6 to 8 inches long. Shallots and leeks are packed 12 bunches to a 15# carton.

Chives, which are thick clumps of tiny underdeveloped pencil-lead thin onions, are used for their tops. Chives are packed 12 pots to a carton.

Parsley is graded No. 1 and is available year-round, with best supplies in October, November, and December. The two types are flat (Italian), which is strong flavored, and curly, which is mild and familiar as a garnish. Good quality parsley is dark green, crisp, and firm. Yellow indicates aging, and black watery areas indicate bruising. It is packed in cartons containing 60 bunches but may be purchased by the bunch from most distributors. It is an excellent source of vitamins A and C.

Parsnips are graded No. 1 and No. 2. They are grown in the northern United States and California and marketed year-round, with the best supply from October through January. They should be small to medium size, fairly free from scraggly rootlets. Large ones may have a woody core. Avoid wilted, flabby roots. Parsnips are packed with 12 20-ounce bags per carton and in 25# and 50# bags. They are a fair source of vitamin A.

Peas include green (English) peas, snow peas (or pea pods), sugar snap peas, and black-eyed peas. They are graded No. 1 and Fancy. Green peas and black-eyed peas should have green, crisp, full pods and should be shelled before cooking. Snow peas have flat, deflated pods that cling tightly to the immature-looking peas inside. Sugar snap peas are similar, but the pods are a bit plumper, and the ends must be snapped and a string along the back removed. Snow peas and sugar snap peas should be bright green and crisp. Both are eaten pod and all. Peas are packed in 10# lugs or 30# tubs or cartons with varying counts.

Peppers, the commonly used term for capsicums, are a family including red pepper, chili pepper, and sweet bell peppers. Paprika and chili powder are two spices made from peppers. Peppers have no botanical relationship to black peppercorns. The heat content of peppers is related to how much capsaicin is in the pepper—the higher the capsaicin content, the hotter the pepper. They are graded Fancy, No. 1, and No. 2 and are classified in two categories: those with mild- or sweet-fleshed fruit and those with hot or pungent-fleshed fruit. The principal variety of mild pepper is Bell, which may be green or red; also included is Pimiento and Anaheim. The winter supply comes from California, Florida, and Texas. The summer supply is grown throughout the United States. Peppers are marketed year-round. They should be bell shaped, thick walled, and firm with a uniform glossy color. A pale color and soft seeds indicate immaturity. Sunken spots are a sign of decay. New technology will increase storage time and retard decay. Packing is done in 28# to 30# bushel cartons with varying counts. Sliced pepper rings are sold ready to serve in 5# plastic bags. Peppers are an excellent source of vitamin C.

The principal varieties of hot peppers are Cayenne and chilies such as Tabasco, Fresno, Poblano, Jalapeno, Serrano, and Pasilla. They are grown in New Mexico and California as well as being imported.

Potatoes are graded No. 1, Extra No. 1, and No. 2. The principal varieties are russet, round red, round white, and long white. Select the kind of potato best suited for its anticipated use. Russets have thick brown skin, white flesh, and an oblong shape. They are used for baking and deep fat frying because they are dry and fluffy when cooked. The principal varieties are Burbank, Norgold, and Centennial. Round potatoes have a high moisture content that accounts for their waxy appearance, making them ideal for boiling, panfrying, and roasting. Round reds have thin rosy red skin. Principal varieties are Red Pontiac, Red McClure, Norland, and Red LaSoda. Round whites have a somewhat thick brown skin and a waxy beige flesh. Popular varieties are Katahdin, Kennebec, Superior, and Irish Cobbler. Long whites are long with a somewhat flattened shape, thin light brown skin, and white flesh. Often called an all-purpose potato, it is acceptable for any use. The principal variety is White Rose.

Potatoes are available year-round because they are easily stored. Most of the crop is harvested September to November. Early or new potatoes are thin skinned because of immaturity and must be handled carefully and used immediately. Potatoes should be consistent in size, firm, smooth, clean, and well shaped. They should not be cut, bruised, wilted, sprouted, or sunburned. They may be waxed to preserve freshness and colored to enhance appearance. Potatoes should be stored in the dark, as exposure to intense light can cause greening of the skin. They should be kept cool but not refrigerated because the cold causes the starch to convert to sugar, which changes the taste of the potato. Potatoes are sold by specific gravity as well as size and grade. The higher the specific gravity, the mealier a potato is, making it better for baking. A potato with a slightly lower specific gravity is waxy and better for boiling. Potatoes with even lower specific gravity are preferred for frying. Potatoes are packed in 50# cartons with counts of 60 to 120 and in bags weighing 50# and 100# with varying counts.

Value-added potato items include raw chopped, chunks, diced, and whole peeled. There are also oven ready and prebaked potatoes, Parisienne potato balls, potato boats, potato skins, preblanched fries cut for steak fries, French fries, and shoestrings. Potatoes are a fair source of vitamin C and contain some vitamin A and a variety of other vitamins and minerals.

Pumpkins are of the same family as squash, with virtually no distinction between them. Almost all the commercial pumpkin supply is from Illinois and is marketed in October. Size and shape have no effect on flavor and quality. Pumpkins are sold individually and are priced singly or by the pound. They are an excellent source of vitamin A and magnesium.

Radishes are graded No. 1 and Commercial. The principal colors are common red, long white, and black. Red radishes may be globular, oval, turnip, oblong, or long. They are plentiful year-round. Regardless of color, the flesh should be white, plump, crisp, and firm, with a biting, pungent flavor. White and black varieties are usually milder than red. Do not accept those that are soft and spongy. Do not wash them until ready to use. Water droplets on the skin may result in black spots and decay. They are packed without tops in bags with varying counts and weights and in bulk containers of up to 40#. They are also available cleaned, sliced, or as rosebuds in 5# cello bags. Easter egg radishes are available in several different colors. Radishes are a good source of vitamin C, potassium, and magnesium.

Rutabagas are graded No. 1 and No. 2. The main variety is yellow fleshed. Most are grown in Canada and are in season between July and April. They are larger than turnips and have an elongated shape and tan skin. They should be smooth-skinned, firm, and heavy. Size is not a quality factor. Do not accept those with punctures, cuts, or decay. They are waxed to retain moisture and packed in 50# bags or cartons with varying counts. Rutabagas are also available in sticks in 5# cello bags. They are a fair source of vitamin A.

Spinach is graded No. 1 and Commercial. It is the most popular of the greens in the United States. It is available in three varieties: Savoy, which is curly; Semisavoy, which is slightly curled; and Flat-Leaf, which is broad-leaved. It is raised in California and Texas. The leaves should be fresh, crisp, and clean with good green color. Do not accept wilted or yellow leaves or overgrown stems. Buy only the amount needed because it deteriorates quickly. If storage is necessary, add ice to increase moisture. It is marketed year-round, but the peak period is in late spring. It is packed in 20# to 22# cartons with varying counts. It is also available snipped and cleaned and ready for use.

Spinach is an excellent source of vitamins A and C, potassium, and magnesium.

Squash is graded No. 1 and No. 2. Florida is the leading source of supply. The principal varieties of hard-shelled mature squash, often called winter squash, are Acorn, Banana, Butternut, Buttercup, Hubbard, Turban, and Spaghetti. It is marketed year-round, the peak period of availability being in late autumn. Winter squash should be heavy for its size. Do not accept squash with cuts, punctures, mold, or sunburn spots. Winter squash is packed in 20#, 50#, and 80# cartons with varying counts but may be purchased individually. Hard-shelled squash is an excellent source of vitamin A.

The principal varieties of soft-shelled immature squash, often called summer squash, are Crookneck, Scallops, Patty Pan, Cocozelle, Yellow Straightneck, and Zucchini. They are used while immature so the seeds and skin may be eaten. Summer squash is marketed year-round, the peak period of availability being in late spring and early summer. It should be well developed, firm, fresh, and glossy. Do not accept stale, overmature, dull, tough squash. Summer squash is packed in 20#, 50#, and 80# cartons with varying counts but may be purchased individually. Zucchini is also available in coins or sticks in 5# cello bags.

Sweet potatoes are graded No. 1, Extra No. 1, and No. 2. The main varieties are Centennial, Georgia Red, Jersey, Nemagold, Goldrush, Puerto Rico, and Velvets. The two types are dry meated and moist meated. The skin and flesh of the dry type are usually light yellow-tan, while the skin and flesh of the moist type vary in color from whitish tan to brownish red. Order thick, chunky, medium-sized sweet potatoes with a uniform tapered shape, firm texture, and bright, clear skin. Do not accept those that are soft or shriveled or with any sign of decay, which spreads rapidly and affects the taste. They are marketed year-round with the largest amounts available in late fall and are packed in cartons with varying weights and counts. They are an excellent source of vitamin A and magnesium and a good source of vitamin C.

Tomatoes are graded No. 1, Combination, No. 2, and No. 3. They are judged on four factors: color or general appearance, firmness, internal appearance, and flavor. Size is a matter of preference. Good quality tomatoes should be full size, ripe, unblemished, and tomato red. An artificially inverted gene has been developed that causes the tomato to soften more slowly and be stored 50 percent longer. Vine-ripened tomatoes are picked mature, ripe, and ready to eat. Vine pink are harvested just as they are beginning to turn. Mature green are harvested when they are mature but still green, giving them a longer shelf life. They are ripened with exposure to ethylene gas in storage. They are usually firmer but have less flavor because they have been refrigerated in transit. Tomatoes grown in the greenhouse or hydroponically often are less flavorful than those that are field grown. Florida and California are the leading domestic producers. Tomatoes are also imported from Mexico and Israel.

Tomatoes are marketed year-round and are packed in 18# to 20# lugs with counts of 50 to 60; in 28# to 30# lugs with counts of 108, 126, and 147; and in 30# cartons with varying counts. Plum or Red Roma tomatoes are small to medium size with an elongated oval shape. They are packed in lugs the same weight as regular tomatoes, but comparable lugs contain up to twice the count because of the small size. The cherry tomato is a miniature variety that is 1 to 1½ inches in diameter when mature. Cherry tomatoes are packed in trays holding 12 pints and weighing 14# to 20#. A garnish, tomato roses are a value-added item that must usually be special ordered. Fresh ripe tomatoes are a good source of vitamins A and C.

Turnips are graded No. 1 and No. 2. They may be either white or yellow with a purple collar. The peak period of availability is late autumn. They should be smooth, round, and firm. Do not accept those with too many leaf scars around the top and fibrous roots. They are usually packed without tops in 25# to 50# bags with varying counts, but early crops may be in 43# to 47# cartons with counts of 24 bunches with tops. Turnip tops are an excellent source of vitamins A and C.

Watercress has long stems and round dark green leaves. It grows in ponds and streams. The pungent flavor is distinctive and peppery in taste. It is fragile, so it should be kept moist in a covered container, refrigerated, and used quickly. Garden cress is similar except it is grown in gardens or greenhouses and has shorter stems.

Table 12.1 lists other less well known vegetables that might be used in some foodservice operations.

Table 12.1. Less well known vegetables used in foodservice operations

Alfalfa sprouts	Fiddlehead ferns
Anise	Ginger root
Arugula	Horseradish root
Bamboo shoots	Jerusalem artichokes (sunchokes)
Bean sprouts	Jicama
Bitter melon	Kohlrabi
Burdock	Mache
Cactus leaves (Nopales)	Mung bean sprouts
Cardoon	Nappa
Celeriac (celery root)	Nopales
Chayote	Red Swiss chard
Chervil	Salsify
Chinese long bean (dow kwok)	Sorrel
Cilantro (Chinese parsley)	Tamatillos
Daikon (Japanese white radish)	Taro root
Dandelion greens	Wheat sprouts
Dill	Winged beans
Fennel	Yucca root

13

CANNED, FROZEN, AND DRIED FRUITS AND VEGETABLES

FRUITS ARE CANNED, FROZEN, AND DRIED so that particular items available in fresh form only during a short growing season and shipped from remote localities may be enjoyed all year. Processing usually changes the structure of the food, reduces the cost, eliminates waste, reduces or eliminates preparation, and extends the storage life. Fruits may be packed separately or in combinations. They may be processed as whole fruits or in halves, slices, sauce, or juice or as jams, jellies, or preserves (see Ch. 16).

CANNED FRUITS AND VEGETABLES

The price of canned fruits and vegetables is affected by the treatment of the product in the canning process. Purchasing decisions should be based on intended use. Cut or broken pieces may be used in some recipes without adversely affecting the finished item and may be sold at a lower cost than perfect pieces. Whole vegetables cost more than pieces or cut vegetables, fancy cuts cost more than diced or short cuts, and vegetables all the same size cost more than mixed sizes. The addition of special seasonings and butter sauce also adds to the cost of the product.

Containers may be described either by a size number or the weight of the contents. Cans made of steel that have been coated inside and out with tin are being replaced by aluminum, which is lighter and has less of an effect on the acidity of the product. Form L at the end of the chapter shows common can sizes, capacities, and uses.

Canned vegetables have approximately the same composition as that of the same weight of the fresh product. Nutrients are lost through processing temperatures and solubility in liquids. Tomatoes, peas, corn, green beans, sweet potatoes, pumpkins, lima beans, and asparagus withstand processing with the least change; the leafy vegetables lose character at the high temperatures required for canning. Green and leafy vegetables change in flavor, texture, and color as well as nutritive value. Cabbage is canned as kraut, and cucumbers are processed as pickles.

Some vegetables are processed before canning. For example, tomatoes may be whole, in quarters, or in pieces or may be made into juice, catsup, paste, puree, or chili sauce.

Canned fruit is graded A, B, and C or Fancy, Choice, and Standard for most fruits. The grade is the most important factor to consider when purchasing canned fruit. Some companies use their own names for these designations, such as Red Label or Blue Label. The following factors are taken into consideration when grading canned fruit: color and uniformity of color, uniformity of size and shape, absence of defects such as frost, insect damage, or bruises, flavor, character (texture, firmness, etc.), Brix range of the syrup (the percentage of sugar by weight in a sugar solution), and drained weight. The USDA has set up point scales on which the relative merits of fruits are based. Assignments of points for various qualities differ with the variety to be graded, but Fancy must have a total of 90 points out of 100, Choice 75 to 89 points, and Standard 60 to 74 points. If there is no B grade, the points for A are 85 or more and for C 70 to 84.

The sizes of fruit are designated by the numerical count according to the size of the can. Usually, the larger the fruit size, the lower the count. Drained weight, the weight of the fruit after the liquid has been removed, is included in the federal standards.

Traditionally, the syrup density is according to the grade, so the greatest amount of sugar is used for the best grades. Cost, and a trend toward less sugar in the diet, are leading to use of a lighter syrup or fruit packed in juice alone. Plain water with artificial sweetener is the medium for dietetic fruits.

FROZEN, DRIED, AND IRRADIATED FRUITS AND VEGETABLES

Frozen fruits and vegetables are packaged in waxed or polycoated paperboard containers or packed in polyethylene, foil, or laminated bags or envelopes. Frozen fruits are also packaged in cans. Frozen fruits and vegetables may be in bulk packs or individually quick frozen.

Dried fruits have been processed to remove moisture and are categorized according to the method used. Dried fruits are called "regular" if 18 to 25 percent of the original moisture remains and are "low moisture" if only 2.5 to 5 percent of the moisture remains. Dried fruits are usually packed in polyethylene bags. Dried vegetables have limited market value.

Dehydrated fruits and vegetables have had even more moisture removed than dried produce.

Freeze-dried fruits and vegetables are first frozen and then dried in a partial vacuum. They have very little color and texture change and may be stored at room temperature. They are quite expensive.

Dehydro-frozen fruits and vegetables are first dehydrated three-quarters of the way and then frozen. Because most of the moisture loss occurs in the last quarter of the drying, partial drying makes for better-textured food. This process reduces bulk, but dehydro-frozen foods must be refrigerated.

Reversible compressed fruits and vegetables are freeze-dried and then compressed into ¾-inch-thick disks. When the foods are put into water, they pop back to their normal size and shape and have the quality of thawed frozen foods. Carrots, green beans, corn, and berries were the first products to be processed in this way.

Irradiated fruits and vegetables are preserved by sterilization with radiation. This causes little change in product quality, and they may be stored without refrigeration.

FORM L

CAN SIZES

Can name	Dimensions		Capacity (approx.)	Use	No./case
	Width	Height			
6 oz	2⅛	3½	¾ C	Frozen juice concentrates, fruit, vegetables, specialty items	24 or 48
8 oz	2¹¹/₁₆	3	1 C	Fruits, vegetables, specialty items	24 or 48
No. 1 picnic	2¹¹/₁₆	4	1¼ C	Condensed soup, fruit, vegetables, meat, fish products	48
12 oz	2¹¹/₁₆	4¾	1½ C	Frozen juice concentrates, fruit, vegetables, specialty items	24
No. 300	3	4⁷/₁₆	1¾ C	Cranberry sauce, spaghetti, chili beans with pork, nut bread	24
No. 303	3³/₁₆	4⅜	2 C	Fruits, vegetables	12 or 24
No. 2	3⁷/₁₆	4⁶/₁₆	2½ C	Fruits, vegetables, juices	12 or 24
No. 2½	4¹/₁₆	4¹¹/₁₆	3½ C	Fruits, vegetables	12 or 24
No. 3 cylinder	5⅛	5⅝	5¾ C (46 oz)	Juices, chicken, soup, grapefruit sections	12
No. 10	6³/₁₆	7	3 qt	Fruits, vegetables	6

14 CEREAL PRODUCTS

GRAINS AND LEGUMES are the dry kernels of plants. The grains are barley, corn, Job's tears, maize, millet, oats, rice, rye, sorghum, teff, triticale, and wheat. The legumes are peanuts, field beans, field peas, cowpeas, soybeans, lima beans, mung beans, chick-peas, pigeon peas, broad beans, and lentils. They are similar in structure but differ in size and shape.

The three structural parts of the kernel are the bran, which is the outer, protective covering of the kernel; the endosperm, which contains the food supply of the plant; and the germ, which contains elements that are necessary for reproduction. The major components of grains and legumes are protein, oil, bran, and starch. Most grains have had the husk or inedible outer portion removed before reaching the market.

Damage is caused to grains and legumes in every form by insects, bacteria, age, and moisture. The process of irradiation will destroy insects and bacteria, proper rotation of a product will prevent unnecessary aging, and good handling practices will control the desired level of moisture, depending on the product.

BARLEY

Scotch barley, sometimes called "barley groats," is unpolished kernels of barley that are sometimes added to salads to provide a chewy texture and a mild, pleasant, nutty flavor.

Pearl barley, the whole grain with hulls and bran removed, is used principally as a soup ingredient. Barley may be processed into barley grits that are ground pearl barley. It may also be made into a flour that is used in baby foods, breakfast cereals, and breads made of a combination of flours. Pearl barley is available in retail packages and 25# bags.

Barley flour is available in retail packages and 10# bags. It is low in protein and fat and a good source of niacin, thiamine, and potassium.

CORN

Corn comes in a spectacular range of varieties and colors, one of the most ancient of which is popping corn.

Cornmeal is made by grinding the cleaned kernels of white or yellow corn to the desired fineness. It contains small amounts of fat and crude fiber and not more than 15 percent moisture according to federal standards. Bolted white or yellow cornmeal is the same as regular, but it is ground finer. Enriched cornmeal contains added thiamine, riboflavin, niacin, and iron. It may also contain additional calcium and vitamin D. Yellow cornmeal has a higher vitamin A content than white. Cornmeal is available in 1#, 5#, 10#, 25#, 50#, and 100# bags.

Corn flour may be a by-product in the preparation of cornmeal or may be prepared especially by milling and sifting yellow and white corn. It is used for the preparation of tortillas and in bread made of a combination of flours. Corn flour is available in retail packages and 10# bags.

Grits are prepared by grinding and sifting white and yellow corn from which the bran and the germ have been removed. They are more coarsely ground than cornmeal and are available in retail packages and 25# bags.

Hominy is corn with the hull and germ removed. Pearl hominy is wholegrain hominy with the hulls removed by machinery. Lye hominy is wholegrain hominy that has been soaked in lye water to remove the hulls. Granulated hominy is a ground form. Hominy is available in retail packages and 25# bags.

Cornstarch is the refined starch obtained from the endosperm of corn. Waxy cornstarch or waxy maize starch is prepared from waxy corn. It acts as a stabilizer for frozen sauces and pie filling. Cornstarch is an ingredient in regular and instant pudding mixes. It is available in 1# packages and 25# bags.

Corn cereals are prepared from corn grits that have been cooked and shaped into flakes, puffs, shreds, and other ready-to-eat forms, then dried or toasted. They may be enriched with thiamine, niacin, and iron. Corn cereals may be combined with other cereals and may be flavored or sugarcoated. They are available in individual portions, retail packages, and bulk packs.

Corn oil is a coproduct of corn processing. It is high in polyunsaturated fats, economical, and bland-tasting, making it a good choice for general use.

MILLET

Millet is the name given to a collection of minor grains grown around the world. The hulled grain has a unique cornlike flavor and firm texture that makes it attractive.

OATS

Oatmeal, or rolled oats, is made by rolling hulled oats to form flakes. Regular oats and quick-cooking oats differ only in the thinness of the flakes. For quick-cooking oats, the groats (the edible portion of the kernel) are cut into tiny particles that are then rolled into thin flakes. Oatmeal is used as a cereal and an ingredient in baking. It is available in retail packages and 50# bags.

RICE

Rice is known to have 7,000 varieties, but usually two are eaten in this country, brown and white.

Brown rice is hulled, meaning the grain has had only the hull removed. The bran is left on the grain; it is not pearled (polished). Brown rice has more nutrients than white rice, a nutty flavor, and a firm texture. It is easier to digest but takes almost twice as long to cook.

White rice is the white, starchy, polished endosperm of rice grains. Polished white rice has the husk, bran, and germ removed, depleting it of fiber, thiamine, protein, and oil. Most white rice is enriched, but the food value never reaches that of brown rice. Rice is classified as long grain, medium grain, and short grain. Enriched rice is prepared by including a percentage of kernels that have been enriched with B vitamins and iron.

Parboiled rice has been steeped in warm or hot water, drained, steamed (usually under pressure), and dried before it is hulled and milled.

Converted rice is parboiled rice made by a specific patented process.

Precooked (instant) rice is packaged long grain rice that has been cooked, rinsed, and dried by flash freezing. This rice requires a very short preparation time. Rice is available in retail packages and in 10# and 25# bags.

Wild rice is the long brownish grain of a reedlike water plant, not a true rice. It is hulled but not milled.

Risotto is a short imported Arborio rice that, when cooked, is creamy rather than the traditional fluffy separate-grain rice.

Aromatics are among the most unusual of the long grain varieties; they exude a light fragrance like that of nuts or popcorn. Aromatic varieties are growing in popularity.

Rice bran is made up of bran and germ with varying quantities of hulls. It is a smooth brownish powder with a faintly sweet taste.

Rice polish consists of inner bran layers and some endosperm. It is a smooth yellowish powder with a sweet taste.

Rice flour is a starchy white flour milled from white rice. Waxy rice flour is made from waxy rice and acts as a stabilizer in sauces and gravies and prevents separation in frozen products. Rice flour and waxy rice flour are available in retail packages and 10# containers.

RYE

Rye flour is the finely ground product obtained by sifting rye meal. It is available in three grades: white, medium, and dark. Rye flour produces gluten of low elasticity, so when used alone, it makes a very compact product. It is usually combined with another kind of flour to make a lighter product. Rye flour is available in retail packages and 50# bags.

TRITICALE

Triticale is a modern hybrid of wheat and rye. It has a higher protein content and better amino acid balance than either of its parent grains. It was originally developed by the USDA to provide a higher-quality diet for those nations who relied on grain for the larger portion of their intake. Whole triticale makes a delicious hot cereal. Triticale sprouts are crisp and nutty flavored.

WHEAT

Wheats are classified as hard (high-gluten) or soft (low-gluten). Hard wheats are used for bread and biscuits, while soft wheats are better for tender cakes and pastries. Wheat may be purchased as raw whole wheat, bulgur, or flour. Some wheat products, such as wheat germ, cracked wheat, and wheat meal, are available in retail packages but are often found only in health food or specialty stores.

Bulgur is the wheat berry that is steamed, dried, and then cracked. It tastes something like wild rice with a mild, nutty flavor. It packs, ships, and stores easily. It may be used as a meat extender in meatloaf, chili, and spaghetti sauce. It adds a nutty flavor to baked goods and fiber to salads.

Bulgur wheat, or parboiled wheat, is whole red or white wheat. White is preferred because of its light golden color. The grain is precooked to partially gelatinize the starch content of the kernel; then it is dried and partly debranned. It may be purchased either as whole grain wheat or as cracked wheat with a fine, medium, or coarse texture.

Rolled wheat is available as a breakfast cereal. It adds a nutlike flavor when added to cookies, breads, meatloaf, and casseroles.

Wurld wheat is similar to bulgur but is much lighter in color as a result of chemical peeling of the bran. Wurld wheat is higher in cost and less nutritious than bulgur.

Wheat germ is the fat-containing portion of the wheat kernel. The germ is flattened and then sifted out as an oily yellow flake. It is high in polyunsaturated fat and vitamin E.

Cracked wheat is prepared by cracking or cutting cleaned wheat, other than durum, into fragments.

Wheat meal is prepared from finely ground whole wheat or parts of the grain.

Ready-to-eat wheat cereals are prepared from whole wheat or parts of the grain. These products are precooked, flavored with malt and sugar, and then dried in various forms such as flakes, puffs, or shreds. They may be sugarcoated. Enriched cereals contain additional vitamins and/or minerals that may not have been present in the whole grain in significant amounts. Restored cereals contain added amounts of thiamine, niacin, and iron to restore them to the levels contained in the original grain. Cereals are available in individual serving containers, retail packages, and quantity bulk packs of various sizes.

Wheat flour may be all-purpose or general-purpose and is prepared by milling and sifting cleaned wheat. It consists primarily of endosperm and may be bleached or unbleached. It may be used for most general cookery purposes. Wheat flour is available in 5#, 10#, 25#, 50#, and 100# bags.

Enriched flour is white flour that contains added thiamine, niacin, riboflavin, and iron. It may also contain calcium and vitamin D. All commercially produced flour is enriched.

Whole wheat flour, or graham flour, is prepared by milling cleaned wheat other than durum and red durum in such a way as to retain the natural proportions of bran, germ, endosperm, and the specific nutrients normally contained in the whole unprocessed grain. Enriched and whole wheat flours are available in 5#, 10#, 25#, 50#, and 100# bags.

Gluten flour is a mixture of wheat flour and gluten, which results in a high protein content.

Self-rising flour is flour to which leavening ingredients and salt have been added in proper proportions for general baking. The leavening ingredient is usually a combination of soda and calcium phosphate. Self-rising and gluten flours are available in 1#, 5#, and 10# bags.

Phosphated flour is available in some parts of the country and is used to make biscuits with sour milk and buttermilk to assure leavening regardless of the acidity of the sour milk.

Bread flours are milled from blends of hard spring and winter wheats or from either of these types alone. They are fairly high in protein and slightly granular to the touch. They may be bleached or unbleached. Bread flours are milled primarily for bakers. They are available in 25#, 50#, and 100# bags.

Cake flours are milled from soft wheats and are the most highly refined flours. The protein content is low and the granulation so uniform and fine that the flours feel soft and satiny. These are used primarily for baked products and are available in retail packages and in 25#, 50#, and 100# bags.

Instant, instantized, instant-blending, or quick-mixing flour is a granular all-purpose flour that blends more readily with liquid than does regular flour. To assure good quality in the final baked product, changes in formulas are needed. Instant flours are available in retail packages.

Pastry flours are usually made of soft wheats, are fairly low in protein, and are finely milled. They are used for making pastries and specialty products, mainly by commercial bakers. They are available in 50# and 100# bags.

Wheat sprouts should be used when only $\frac{1}{16}$-inch long. They are high in protein, vitamin E, and niacin and are very sweet. Add them dried to cookie dough and cake batter.

PASTA

"Pasta," the generic word for macaroni, spaghetti, and egg noodles, derives from the Italian phrase "pasta alimentare." Alimentare relates to the function of nutrition, and pasta is the dough made by mixing durum flour with water.

Durum is the very hard, amber-colored, high-protein wheat that is grown in North Dakota. Of the wheat grown in the United States, durum produces the best flavor, color, and texture in pasta. The coarsely ground endosperm of durum wheat is known as semolina. It is enriched with B vitamins and iron and mixed with water to produce pasta products.

Just as semolina is a rough granulation of durum wheat, farina is a rough granulation of any other high-quality hard wheat. It is enriched and can be used in pasta products, blended with semolina, or cooked as a wheat cereal. Enriched farina contains added thiamine, niacin, riboflavin, and iron. Calcium and vitamin D may also be added.

Pasta is often divided into four groups: long goods—spaghetti, linguine, vermicelli; short goods—macaroni, rigatoni, ziti; specialty—manicotti, lasagne, shells; and noodles—pasta made with eggs. Pasta is available in many other shapes than the examples given.

Pasta for foodservice is usually packed in bulk in boxes. Long and short goods are packed in 20# boxes. Egg noodles and specialty products are packed in 10# boxes. Uncooked pasta will keep indefinitely if stored in a cool, dry place.

Couscous is tiny pellets, made from semolina mixed with water and a fine coating of flour. It is steamed until fluffy and has a mild flavor and soft texture.

Egg noodles are a fine granulation of durum wheat with added B vitamins and iron mixed with 5.5 percent by weight of egg solids. It is rolled into thin sheets and cut in strips of varying widths, or it is extruded into its characteristic flat shape. Flavored noodles are made by combining various kinds of dried ground vegetables. Spinach and tomato are the most common pastas, but more exotic varieties are made with artichokes, beets, carrots, red chilies, parsley, basil, or broccoli.

Specialty noodles may be made of mung beans, rice paper, soy, yams, or buckwheat.

Lupini pasta is made from a bean that has been hydrolyzed from a nutritious, rather unpalatable, legume into a tasty, sweet variety. It has been combined with triticale into a pasta that is low in calories and high in protein, calcium, and iron.

AMARANTH

Amaranth is an ancient grain that is making a comeback because of its excellent nutritional qualities. The flour made from amaranth is available in health food stores.

BUCKWHEAT

Buckwheat has a strong, earthy flavor. It contains all eight essential amino acids, vitamin E, B-complex vitamins, and calcium. It has a relatively high fat content, so it shouldn't be stored a long time, or it will become stale. Roasted, buckwheat makes kasha, a ricelike dish. The groats (whole grain) are used in cereal and pilafs.

Buckwheat flour is a finely ground product obtained by sifting buckwheat meal. It is available in retail packages and in 50# bags.

Soba noodles are made with buckwheat. Regular soba is made of buckwheat and regular flour, Yamaimo soba has Japanese mountain yam added to it, and there is a soba noodle flavored with powdered green tea. Soba noodles have a firm, slightly coarse texture.

QUINOA

Quinoa (pronounced *KEEN-wa*) is called a supergrain for its high protein content. Quinoa seed can be added to soups, puddings, and stir-fries. It is not grown in this country and so may be available only by special order.

SOYBEANS

Soybeans are the best known of the legumes. They are very high in protein, making them an important food source. Research is ongoing for ways soybean components can be used in food processing. Some people enjoy roasted soybeans as a snack, but most soybeans are milled and processed.

Soy flour is highly flavored. It is usually combined with wheat flour in baked products because it lacks gluten-forming proteins needed for bread making. Full-fat soy flour is made by grinding soybeans that have only the hull removed. Low-fat soy flour is made from the press cake after all or nearly all the oil is taken out of the soybeans. The soybeans may be heat treated or conditioned with steam prior to oil extraction. Soy flour is available in retail packages and 10# bags.

Soy oil is a coproduct of processing soybeans. It is high in polyunsaturated fats and low in cost. It is usually found blended with other oils or used in food processing.

Soy grits are made from coarsely ground soy press cake and are a low-fat product. Grits are sold in retail packages and 10# bags.

Soy, originally called "textured vegetable protein" (TVP), contained 50 percent protein. It was improved and renamed "vegetable protein product" (VPP). The protein content was increased a little, it was no longer texturized, but it was formulated in a way that eliminated the chewiness that many people found undesirable. A newer product called "isolated soy protein" is 90 percent protein. With care, up to 50 percent replacement of casein/caseinate blends is possible in several cheeses. Quality characteristics such as flavor, melt, and stretch are maintained in the cheeses while fat, sodium, and cost are greatly reduced. VPP is used to make meat analogs that are formed to look like their real-meat counterparts. It may be purchased already blended with ground beef in beef patties, meatloaf mix, and other products, and it may be purchased separately to be added as an ingredient in a variety of recipes.

Soymilk is made into tofu (sometimes called "soy cheese" or "bean curd") by a process similar to making cheese. The soymilk is coagulated by using calcium sulfate or a curding agent made from seawater called "nigari." The curds are pressed into soft cakes, resulting in a soft custardlike food that is an excellent source of protein. Tofu has a mild odor and usually a bland taste; however, the flavor varies depending on the type of soybean used in processing. It readily absorbs the flavors of the seasonings used with it. Tofu comes in varying textures and degrees of firmness. Intended use determines the type to choose. It is sold water packed in plastic tubs, vacuum-packed, or aseptically packaged. Aseptic packs are airtight cardboard boxes that don't need refrigeration until they are opened. Avoid tofu that isn't white or that is stored in cloudy water. Soymilk is also being used in donuts to reduce their fat absorption.

Soybean sprouts have been used in Oriental dishes since 3000 B.C. They are high in protein and vitamin B.

VEGETABLE OILS

Vegetable oils are recommended as a substitute for butter in some diets as a way of lowering serum cholesterol. Not all oils accomplish this, even if they are processed from grains. In addition to corn and soy oil, already mentioned, vegetable or nonanimal fats include coconut, palm, palm kernel, cottonseed, olive, peanut, sunflower, safflower, and canola, which is made from rapeseed. Researchers are studying the effects of fat on health, so choices of fats and oils will need to change with new information.

15 SPICES, HERBS, AND FLAVORINGS

SPICES AND HERBS are aromatic natural products used to aid and enhance the natural flavors of foods, not to obscure them. They may be the root or rhizome, the bud, the seed, the bark, the fruit, or the flower. There is a difference of opinion among experts as to the classification of herbs and spices. This compilation will be divided into ground spices, herbs, seeds, blends, and vegetables.

Seasonings should be stored in a cool, dry place, since heat reduces their flavor and moisture will cause them to cake. The other concern with storing spices is insects. These are destroyed by irradiation, but currently only a small fraction of the herbs and spices in this country are irradiated.

SPICES

Ground spices give up their flavors quickly and are used in uncooked dishes or added near the end of the cooking period. Whole spices are added at the beginning of a long cooking period and removed before serving. The following whole or ground spices may be used in food service:

Allspice is a pea-sized fruit that grows in small clusters on a tree. Picked green, the fruit appears as shriveled brown berries after curing. It is available whole or ground.

Capsicum pepper is a dried, many-seeded berry. There are many varieties, including cayenne and chili pepper. It is ground to yield a hot reddish powder.

Cinnamon comes from the bark of an aromatic evergreen tree. Cassia is reddish brown and has pungently sweet aroma and flavor. Ceylon cinnamon is more buff colored and has a milder flavor. It is available whole or ground.

Cloves are the fruit of a tree belonging to the evergreen family. They are a dark brown, dusky red color. The flavor is a sweet, pungent spiciness. It is available whole or ground.

Garlic is a member of the lily family. It is a bulbous plant with a papery sheath covering small white or purplish cloves.

Ginger is the rhizome of a tuberous plant and has a weak yellow-orange skin and yellow-brown interior. Its flavor is warm with a pungent spiciness. It is available fresh, whole, preserved (candied), or ground.

Mace is the fleshy growth between the nutmeg shell and the outer husk. It is orange and has a flavor that resembles nutmeg. It is available whole or ground.

Nutmeg is the kernel of the nutmeg fruit, which grows on a tree. The fleshy husk splits on one side, releasing the aromatic deep brown nutmeg. The flavor is sweet, warm, and spicy. It is available whole or ground.

Paprika is a sweet red pepper that is ground after the seeds and stems have been removed. It is mild and sweet in flavor, slightly aromatic, and a brilliant red.

Pepper is the small dried berry of a vine. Whole pepper berries are known as peppercorns. White pepper is the same berry with the outer black cover removed. The flavor of both black and white pepper is warm and pungent. Black pepper is available whole, ground, or coarse ground. White pepper can be whole or ground. Both are sometimes mixed with other flavorings. Pepper is the most commonly used spice.

Saffron is the world's most expensive spice. It is the stamen of a crocuslike flower. The flavor is distinctive and agreeable and gives food an appetizing yellow color.

Turmeric is a root in the ginger family. It is bright orange-yellow and has a flavor that is mild ginger-pepper. It is an important ingredient in curry powder and is used as a seasoning for pickles. It is available ground.

HERBS

Herbs are generally the aromatic leaves and sometimes the flowers of plants, although other parts are also used. They may be purchased and used fresh, but because of cost and short storage time, it is often more practical to purchase them dried. Usually, a small amount is all that is needed to provide characteristic flavor. Herbs retain flavor longer if stored as whole leaves. The following herbs may be used in foodservice:

Basil, also known as sweet basil, comes from the cleaned, dried leaves and tender stems of a bushy plant that is grown domestically. It has a pleasing licorice flavor. The leaves are available crushed or powdered.

Bay leaves are the dried leaves of the laurel tree, an evergreen. The glossy, smooth, oblong leaves are deep green on the upper surface and paler beneath. The flavor is sweet and herbaceous with a delicate floral spice quality. It is available in leaf form.

Chervil is an aromatic herb of the carrot family. The delicate leaves look and taste like parsley with a hint of anise. It is available crushed or powdered.

Chives are long slender blades with a pleasant oniony flavor.

Coriander is also called "cilantro" by the Spanish or may be called "Chinese parsley." Its leaves look like parsley but have a cool, citrus, minty flavor. It is available as leaves or ground seeds.

Dill weed is a European herb of the carrot family with fine feathery leaves that have a caraway taste. It is available crushed or powdered.

Horseradish is a tall, coarse, white-flowered herb of the mustard family. The pungent root is ground for use as a bottled condiment. It is also available dried and powdered.

Marjoram is an herb of the mint family grown domestically. It has a peculiar sweet minty flavor. The leaves are available crushed or powdered.

Mint is grown domestically. Spearmint has large pointed green leaves and a light minty taste. Peppermint has round fuzzy leaves and a strong, sweet flavor. Both are available as dehydrated flaked leaves.

Oregano is a variety of sage with small fuzzy leaves that have a bittersweet, spicy taste. It is available as crushed or ground leaves.

Rosemary is a spiky herb that looks like a curved pine needle. It is sweet and fresh tasting. It is available as leaves or ground.

Sage is a perennial shrub grown in the United States. The long velvety leaves have a potent, warm muskiness. The flavor is camphoraceous, with a minty spiciness. The leaves are available rubbed or ground.

Savory is an herb of the mint family. The feathery green leaves and blossoms have a flavor that is deliciously sweet. The leaves are available crushed or ground.

Tarragon is a small perennial plant, which forms tall stalks. Grown in the United States, it has leaves that are long, slender, and dark green with a minty, aniselike flavor. It is available as leaves or ground.

Thyme is a low shrub with small pointed leaves. The leaves and stems of this garden herb have a strong, distinctive, pleasantly pungent flavor. It is available as leaves or ground.

SEEDS

Seeds are the aromatic, dried, small whole fruit or seeds of plants. They may be spices or herbs. The following seeds may be used in foodservice:

Anise is a dried greenish brown seed. It is much used in flavoring licorice.

Caraway seeds are somewhat curved, tapering toward both ends. The flavor is a combination of dill and anise with a slight fruitiness.

Cardamom seeds are tiny brown seeds that grow enclosed in a white or green pod, 0.25 inch to 1 inch long. The flavor is sweet and spicy with a camphoraceous note. It is available whole or ground.

Celery seed is a minute olive-brown seed from the celery plant. It has a parsley-nutmeg flavor. Celery salt is a combination of ground celery seed and salt.

Coriander is the dried fruit of a small plant. Externally, it is weak yellow-orange to moderate yellow-brown, often with a purplish red blush. The flavor has a sweet, dry, musty spice character tending toward lavender. It is available as seed or ground.

Cumin is a small dried fruit, oblong in shape. It resembles caraway seeds. The flavor is penetrating. It is an ingredient in chili powder and is available as seed or ground.

Dill seed is the small dark seed of the dill plant that is grown in the United States. It is brown, broadly oval, and rounded at both ends. The flavor is clean with a green, weedy character.

Fennel is related to the carrot family and has a small seedlike fruit. It is agreeably aromatic with a sweet taste similar to anise. It is grown domestically.

Fenugreek is the aromatic seed of an annual Asiatic herb of the pea family. It is used in making curry powder.

Mustard is a small seed. Brown or black mustard has dark brown round seeds; yellow or white mustard has oval seeds and is milder flavored. The brown and black is the pungent variety. All are domestically grown and are available as seed or powdered.

Poppy seeds come from capsular fruit of bristly haired plants characterized by showy flowers. The tiny round seeds are black when ripe.

Sesame is a small honey-colored seed with a gentle nutlike flavor. It is produced in the United States and is available as hulled seed.

Poppy and sesame seeds are usually toasted before they are used. Because of their high oil content, they are never sold in the ground form.

BLENDS

Many blends have been developed by spice manufacturers to make seasoning easy. Some of the more popular combinations used in foodservice follow:

Apple pie spice is a ground blend of cinnamon with small amounts of cloves, nutmeg or mace, allspice, and ginger.

Barbecue spice is a ground blend of chili peppers, cumin, garlic cloves, paprika, salt, and sugar.

Chili powder is a ground blend of chili peppers, oregano, cumin, garlic, salt, and sometimes cloves, red pepper, and allspice.

Crab boil is a combination of whole spices and herbs to be added to water when boiling seafood. It includes whole peppercorns, bay leaves, red peppers, mustard seeds, ginger, and other spices.

Curry powder is a ground blend of 16 to 20 spices and herbs. Most blends will include ginger, turmeric, fenugreek, cloves, cinnamon, cumin, coriander, black pepper, red pepper, and others according to the preference of the blender.

Herb seasoning is a savory blend of herbs often including basil, marjoram, oregano, chervil, etc.

Italian seasoning is a blend of Italian spices and herbs such as oregano, basil, red pepper, and rosemary. Garlic powder and other flavorings may also be included.

Mixed pickling spice is a mixture of whole spices and herbs, usually including mustard seed, bay leaves, black and white peppercorns, dill seed, red peppers, ginger, cinnamon, mace, allspice, and coriander seed.

Poultry seasoning is a ground blend of sage, thyme, marjoram, savory, and sometimes rosemary and other herbs.

Pumpkin pie spice is a ground blend of cinnamon, cloves, and ginger.

Seafood seasoning is a ground blend of the spices and herbs used in crab boil and shrimp spice with the addition of salt.

Seasoned or flavored salt is a mixture of spices, herbs, and salt designed to be an all-purpose seasoning.

Shrimp spice is a combination of whole spices and herbs similar to crab boil.

DEHYDRATED VEGETABLE SEASONINGS

Dehydrated vegetable seasonings are intended to be laborsaving. They rehydrate in 5 to 20 minutes depending on the product. They may be rehydrated as part of the cooking process or separately to be used as a garnish or ingredient. Most of the products are grown in the United States, with California as the largest supplier.

Celery flakes are dehydrated flaked leaves and stalks of celery.

Chives are dehydrated and minced.

Garlic may be dehydrated and minced, powdered, or mixed with salt.

Mixed vegetable flakes are a mixture of dehydrated flaked celery, green peppers, and carrots.

Onions may be dehydrated and sliced, chopped, minced, powdered, or mixed with salt.

Parsley flakes are dehydrated flaked parsley leaf and stem material.

Pepper flakes or sweet pepper flakes are dehydrated flaked sweet green or red peppers or a mixture of the two.

FLAVORINGS AND FOOD COLORINGS

Flavorings are liquid additives made by combining the oil or essence of a product with alcohol and water. The most common of these are vanilla and peppermint. Others include almond, anise, brandy, mint, orange, and wintergreen. Artificial flavorings are prepared by combining selected oils and esters with alcohol and water. They include banana, black walnut, burnt sugar, butter, chocolate, lemon, maple, rum, strawberry, and vanilla.

Food colorings, although not flavorings, are used to change or enhance the color of the food to which they are added. They are available in liquid, paste, and powdered form. The most popular colors are red, blue, black, brown, flesh, orange, peach, pink, green, yellow, caramel, and egg shell.

16 SWEETENING AGENTS

THERE IS A WIDE VARIETY of sweetening agents used as additives to make foods and beverages taste sweeter. Most of the caloric sweetening agents are sugars and syrups. Sugars are carbohydrates. They range from monosaccharides, which are single molecules, and disaccharides, composed of pairs of monosaccharides, to polysaccharides, composed of combinations of mono- and disaccharides. Fructose and dextrose, also called glucose, are monosaccharides. Maltose, lactose, and sucrose are disaccharides. All other complex sugars and starches are polysaccharides.

SUGARS

Sugar is refined sucrose extracted from sugar beets and sugarcane. Sugars from these two sources are both 99.5 percent pure sucrose. White granulated sugar is the standard product for general use. Manufacturers may have variations in particle size and names for the variations; these do not affect the sweetening ability. Sugar is available in numerous sizes and types of packages, from 1# cartons to 100# bags, as well as individual packets containing approximately 1 teaspoon.

Superfine granulated sugar is uniformly fine grained. It is designed for special use in cakes and mixed drinks and other uses where rapid dissolving is desirable. It is available in 1# cartons.

Powdered or confectioner's sugar is granulated sugar that has been crushed and screened to a desired fineness. It is used in icings and frostings and for dusting doughnuts and pastries. It is usually combined with a small amount of cornstarch to prevent caking. It is available in 1# cartons and 2#, 25#, and 50# bags.

Cut tablets are made from sugar that has been molded into slabs that are cut or clipped into various sizes and shapes. Tablets are packed in 1# and 2# cartons.

Pressed tablets are made by compressing moist white sugar into molds to form the various shapes, which are oven dried to produce hard, smooth units of various sizes.

Cubes are also formed in molds. Sizes range from 200 to 80 pieces per pound. Cubes are packed in 1# and 2# cartons.

Brown sugar contains molasses, nonsugars naturally present in molasses (ash), and moisture in varying ratios. It may be produced from the syrup remaining after the removal of commercially extractable white sugar or by the addition of refined syrups to specially graded, uniformly minute white sugar crystals. It may be yellow, golden yellow, golden brown, light brown, or dark or old-fashioned brown. The strength of molasses flavor increases with color intensity. It may be obtained in moist or dry granulated form. Care should be taken to avoid drying in storage. Brown sugar imparts a pleasing, characteristic flavor to baked goods and candy. It is available in 1# cartons and 2#, 25#, and 50# bags.

Sucanat is a brown sugar made from evaporated sugar cane juice. It is very costly and usually sold in health food stores.

Corn sugar is crystallized dextrose (glucose) obtained by hydrolyzing cornstarch with acid. It is used commercially in a wide variety of familiar food products. It is hygroscopic (attracts moisture readily) and so must be handled and stored very carefully to avoid caking.

Raw sugar is unrefined crystalline sugar. The crystals are dark, coarse, and sticky because they contain the molasses portion of the sugar.

Lactose is milk sugar used in food processing. It is a disaccharide composed of glucose and galactose and is digested by the enzyme lactase.

Maltose, or malt sugar, is a disaccharide of two glucose molecules. It is created during bread making and brewing and is commercially produced by breaking down starch.

Maple sugar is the solid product resulting from evaporation of maple sap or maple syrup. It consists mostly of sucrose with some invert sugars (dextrose and levulose) and ash.

SYRUPS

Cane syrup is the sap of sugarcane, concentrated by evaporation. The recommended maximum ash content is 4.5 percent in the unsulfured product and 6 percent in the sulfured product. It is used primarily as an ingredient in other syrups.

Molasses is the by-product of sugar manufacture. Table molasses is light in color and contains a higher percentage of sugars and a smaller percentage of ash than cooking molasses. Cooking molasses is dark (blackstrap). Barbados molasses is a specially treated cooking molasses that resembles cane syrup more than molasses in composition. Molasses is available in retail, ½-gallon, and 1-gallon containers.

Refiners' syrup is the residual product obtained in the process of refining raw cane or beet sugar. It is a solution and suspension of sucrose, partially inverted sucrose, and not more than 28 percent moisture. It is clarified and decolorized and used extensively in flavoring corn syrup.

Sorghum syrup is obtained by concentration of the juice from sorghum cane. It has a distinctive flavor and may be used as a substitute for molasses and other syrups. It contains not more than 30 percent water or 6.25 percent ash calculated on a dry basis. It is available in retail containers.

Corn syrup is obtained through one of several methods of partially hydrolyzing cornstarch. The liquid is neutralized, clarified, and concentrated to the desired consistency. The principal sugars are dextrose, maltose, and dextrins. Light corn syrup has had clarifying and decolorizing treatment. Dark corn syrup is a mixture of corn syrup and refiners' syrup, which gives a darker color and distinctive flavor. Corn syrup is available in retail and 1-gallon containers.

Fructose is a sweetener made from corn. Through a process called "isomerization," the atoms in the dextrose molecule of cornstarch are rearranged to produce fructose. High-fructose corn syrups (HFCS) are very popular in the food industry. The largest user is the soft-drink industry, followed by the canning, dairy, and baking industries.

HFCS is available in three concentrations: 42 percent HFCS is made of 42 percent fructose, 52 percent dextrose, and 6 percent polysaccharides; 55 percent HFCS is made of 55 percent fructose, 40 percent dextrose, and 5 percent polysaccharides; 90 percent HFCS is made of 90 percent fructose, 7 percent dextrose, and 3 percent polysaccharides. Fructose is generally considered the sweetest of all sugars, but the perceived sweetness will vary, depending on how it is used. When fructose is used in warm foods, its sweetness is similar to table sugar, but in cold foods, the perceived sweetness is 1.5 times that of table sugar. Fructose is superior to sucrose as a browning agent; it enhances natural flavors, may be stored at room temperature without deterioration, and is economical.

HFCS is available in retail, 1-gallon, and 55-gallon containers. A crystalline form of fructose is also available in individual portions, retail packages, and quantities by special order through the supplier.

Maple syrup is made by evaporation of maple sap or by solution of maple sugar. It contains not more than 35 percent water and weighs not less than 11# per gallon. It is available in individual portions and a variety of tabletop sizes and in 1-gallon containers. It is very expensive. Many facilities choose to serve imitation maple syrup, which may be prepared by combining a variety of other syrups with maple flavoring or imitation maple flavoring.

Honey is the nectar of plants that has been gathered, modified, stored, and concentrated by honeybees. The top grade of honey is U.S.

Grade A or U.S. Fancy. Less desirable honey is U.S. Grade B. The most important factors in the grading of honey are flavor, clarity, and absence of defects. The different flavors of honey are classified according to the plant from which the nectar was taken. The mildest honey is from clover. The principal ingredients in honey are levulose (fructose) and dextrose (glucose). The water content is limited to 20 percent. Crystals may form in stored honey because of the concentration of the sugar solution, but they can be dissolved by heating. Honey is available in individual portions, retail containers, and 5# and 6# containers.

Most of the honey on the market is extracted honey or honey that has been separated from the comb. Honey creme or spread has been commercially crystallized to make it a more spreadable product. Comb honey is expensive because it is difficult to produce and requires special care in handling and storage. Honey creme, honey spread, and comb honey are available in retail containers.

JELLIES, JAMS, PRESERVES, AND MARMALADES

Jellies, jams, preserves, and marmalades are fruits or fruit juices that have been cooked with sugar to the desired consistency. The FDA has established standards of identity for jams, jellies, and preserves. To meet the standards, the composition must be no less than 45 parts of fruit to 55 parts of sweeteners by weight. Products not meeting the standards must be labeled "imitation." Jelly is a clear product made of fruit juice with no fruit solids. Jam is a product made of juice and pieces of fruit that have not retained their original shape. Preserves is a product made of fruit that has maintained its original shape throughout preparation. Marmalade is a thick, pulpy jam, usually containing shreds of fruit rind. All four products are available in individual portions, retail containers, and No. 10 cans.

Jams and jellies that will withstand the high temperatures of baking are available from bakery suppliers for use in sweet rolls, coffee cakes, and other fruit-filled pastries. They are made with waxy maize starch for clarity and to prevent breakdown at high and low temperatures. They are available in 2# pouches and 40# pails. Marmalade comes in a 20# pail.

Jams and jellies with less viscosity are used as toppings for ice cream and other desserts. They are available in individual portions, retail containers, No. 10 cans, and 30# pails.

POLYALCOHOLS

Sorbitol, mannitol, and xylitol are three polyalcohols or sugar alcohols serving as sweeteners. Sorbitol and mannitol are about 0.5 to 0.7 times as sweet as sucrose. Sorbitol is used in dietetic candies and chewing gum. Mannitol is used as a bulking agent in powdered foods and to dust chewing gum. Xylitol has the same sweetness as sucrose and is used in various foods and in chewing gum.

NONNUTRITIVE SWEETENERS

Nonnutritive (or extremely low-calorie) sweeteners are used in low- or no-calorie beverages, dietetic canned fruits, confections, dairy products, and baked products as well as in individual service packets. Also known as "artificial sweeteners," they are available in both liquid and crystalline form where they are approved for use.

Saccharin is a noncaloric (not metabolized) sweetener that is 300 times sweeter than sucrose. It is colorless, odorless, and water soluble and has a slight aftertaste. It has a stable shelf life, combines well with other sweeteners, and has a synergistic effect when combined with other sweeteners (the combinations are sweeter than the sum of the individual sweeteners).

Aspartame (Equal or NutraSweet) is a dipeptide made from two commercially produced amino acids, L-phenylalanine and L-aspartic acid, which are identical to those found naturally in food. The nutritive sweetener supplies four calories per gram when metabolized. It is 200 times sweeter than sucrose and has a sugarlike taste and a synergistic effect when combined with other sweeteners. It is unstable at prolonged high heat and so cannot be used for baking, and it breaks down gradually in liquid.

Cyclamate is a noncaloric sweetener that is 30 times sweeter than sucrose. It has a stable shelf life, is soluble in liquids, has a sugarlike taste, and has a synergistic effect when combined with other sweeteners. It has no aftertaste and is cold and heat stable. It was banned in the United States as a possible cancer-causing agent in animal studies. Later studies show no cancer link in humans.

Acesulfame K (Sunette or Sweet One) is a nonnutritive sweetener that is 200 times sweeter than sucrose. It is an organic salt containing potassium (K). The sweet taste is rapidly perceptible; it has good shelf life and is relatively stable across normal temperatures, including baking temperature ranges and pH ranges. It has an aftertaste in some products and is not metabolized.

Other products being tested are thaumatin, stevioside, alitame, sucralose, chloroderivatives of sucrose, dihydrochalcones, and the L-sugars.

17 BEVERAGES

COFFEE

Coffee is grown near the equator. The yield varies, but on the average, one tree will produce 1# of coffee annually during its productive years. About 3,500 beans are required to make up 1# of roasted coffee. There are about 50 varieties, which fall into two general classifications: arabica and robusta, each with a characteristic shape, size, color, flavor, and aroma.

The robusta variety, grown mainly in Africa, India, and Indonesia, tends to have a harsh flavor. Robustas are used mainly in instant coffees and in blends. They have about twice the caffeine content of arabicas. Many decaffeinated coffees are made from robusta beans, since the processing rids the beans of much of their natural harshness while providing a large harvest of caffeine for separate sale to makers of soft drinks, drug products, and other products.

Most beans found in wholesale markets are arabicas. There are two main kinds, brazils and milds. Brazils comprise any Brazil-grown beans. There are dozens of varieties, but the main ones take their names from the ports of export. Santos beans have a sweet, clear flavor. Rios are pungent. Other important kinds are Victorias and Paranas. Milds are those arabica coffees not grown in Brazil. They come from Central and South America, the Caribbean, the Middle and Far East, and parts of Africa. These varieties also bear the name of the locale where they are grown. Milds include the following.

Arabian. Only Arabian-grown coffees bear the designation "Mocha." A fine Mocha is heavy bodied with a smooth and delicious flavor and an unusual acid character.

Columbian. Experts consider these winy in body and fine in flavor and aroma. Colombian coffees include Supremo, which is rich, mellow, full-bodied, and mild in aroma; Colombian Espresso, a deep dark roast like Italian Espresso; and Colombian Amaretto, Cinnamon, and Swiss Chocolate Almond, varieties of flavored coffee.

Haitian. These coffees are mellow and rich, with a heavy body and fairly high acidity.

Hawaiian. The principal variety is Kona, with excellent quality, mild acidity, distinct sharpness, and medium body.

Indonesian. Java comes from the island of Java (some robusta is also grown there), and Mandheling is from Sumatra's west coast. These are among the finest coffees in the world.

Tanzanian. Tanzanian Kilimanjaro is grown on the southern slopes of the Kilimanjaro range. It is mild, aromatic, and rich, with a distinct sharpness. Some robustas are grown there also.

When the coffee fruit matures on the tree, it closely resembles a cherry. These cherries are handpicked as they ripen and then are processed for shipment. Each cherry contains two flat-faced beans (seeds) facing each other. The beans are protected by three external layers of skin: the silver skin on the inside, the parchment on the outside, and surrounding these two, the outer skin of the coffee cherry.

Green coffee can be processed to remove the hulls by the wet or dry method. Later, the beans are run through sieving machines that sort them into different sizes. The beans are graded according to the number of imperfections.

After grading, expert "cuppers" check the product to determine its flavor quality. They roast, grind, and prepare a sample and smell the aroma. After it has cooled somewhat, it is tasted. The coffee taster appraises the green coffee and then proceeds to build a blend that will meet the established standards of the taster's company.

Most dealers offer a choice of whole or ground beans and may offer several kinds of roasts. American roast is relatively light with a mild flavor. Continental roast, sometimes called "European" or "New Orleans," is slightly darker and heavier. The French roast makes a still darker coffee. The almost burnt Italian roast makes a rich, somewhat bitter, espresso-type brew.

Commercial blends are made up of numerous varieties of coffee. It is the task of the expert blender to maintain a uniform blend flavor at all times, even though the varieties composing the blend may undergo constant change. Coffee drinkers usually prefer the blends containing less robusta coffee and having less cereal character. They also prefer regular coffee over instant. In some parts of the country, people prefer coffee that is blended with chicory.

Most roasters prefer to blend the green coffee before sending it to the roaster. When roasting time is completed, the process is stopped quickly, usually by means of a cold water spray, and then the beans are cooled to the desired temperature by forced air. The coffee is cleaned again and foreign materials removed.

The roasted coffee is then ground to the desired fineness, measured by means of sieves. After grinding, the coffee goes to the packing area, where the selection of a container will depend on the time lapse before the consumer is expected to use it.

Coffee begins to stale and lose its flavor, strength, and aroma as soon as it is roasted. The rate speeds up considerably after grinding. Ground coffee will stale twice as rapidly as bean coffee. Conditions of distribution and storage in the foodservice industry have led many suppliers to convert to gasflushed packaging, with a shelf life of 13 to 17 weeks, and foil packaging, with a shelf life of 26 weeks. Heat and moisture cause coffee to stale rapidly. The special packaging can protect coffee from moisture but not from heat.

For good results the grind designed for the foodservice's brewing equipment must be used. The grind determines the length of time coffee and water should be together. Too fine a grind produces astringent, bitter coffee and too much sediment. Too coarse a grind results in a weak, flavorless beverage.

Although sold by weight, coffee is measured by volume. Different coffees weigh different amounts; some differ by as much as 40 percent. A relatively lightweight, fluffy coffee, achieved by short roasting, will deliver more cups per pound than a heavier one. Weight differences alone can influence cost per cup.

Decaffeinated coffee is prepared by steaming and soaking the green coffee by using one of several methods. The beans are then combined with a substance that has a special ability to remove caffeine. The common substances used are coffee oils extracted from coffee beans, pretreated charcoal, carbon dioxide, ethyl acetate, methylene chloride (dichloromethane), or water in conjunction with other decaffeinating substances. The beans are then roasted and ground in the same way as regular coffee.

Instant coffee and freeze-dried instant coffee are prepared by freeze-drying and/or by various extraction, evaporation, and drying processes. Decaffeinated instant or decaffeinated freeze-dried instant coffee may also be prepared by these same procedures.

Specialty coffees such as latte, espresso, and cappuccino served both hot and iced are very popular at coffee bars.

TEA

Tea is the name of an evergreen shrub or small tree and the beverage brewed from its leaves. Tea is a favorite beverage that is consumed both hot and iced. The chemical composition of the tea leaf is influenced by soil, genetic variation, climate, maturity of leaves, manufacturing processes, and storage conditions. Tea leaves contain a number of enzymes that cause them to ferment when crushed and exposed to air.

Tea is usually prepared for consumption by brewing. According to the package directions of most commercial products, tea should be prepared by pouring boiling water over tea leaves or a tea bag, then steeping from three to five minutes. However, conditions for tea brewing such as amount of leaves used, temperature of the water, and size of the teacup are subject to considerable variation. The amount of leaves packed in tea bags varies among different brands.

Major types of leaf tea are black, green, and oolong.

Black tea makes up 98 percent of the leaf tea sold in the United states. In the preparation of black tea, the leaves are picked and allowed to dry in air for some time, after which they are rolled and crushed by hand or machine and spread out in air for a longer time than before, this time to ferment. When fermentation is completed, the leaves are rolled again, dried by heat until crisp, and graded. The grade of the tea is related to the size of the leaf rather than the quality.

Green tea differs from black tea by being steamed after picking to inhibit the action of the fermenting enzymes. The grades include the pan-fired and basket-fired green teas of Japan and hyson and gunpowder teas of China.

Oolong tea is partially fermented and then steamed and is therefore intermediate in characteristics between black and green tea. The best grade is Formosa.

Iced tea may be made by brewing tea and adding ice or by putting a tea bag (sometimes a special blend) in tap water and extracting it by placing it in the sun or in the refrigerator.

Instant tea makes up almost half the tea sales in the United States. It is prepared by removing the moisture from brewed tea, usually black. It is available as 100 percent instant tea, sugar-free iced tea mix, lemon-flavored and sugar tea mix, lemon-flavored tea mix, and sweetened tea mix. These products are available in 2-ounce and 3-ounce jars.

Herb tea may be made from black tea to which spices and flavorings have been added during the processing operation. Cinnamon, lemon, orange, mint, almond, clove, apple, and rose hips are some of the many flavorings that may be added. They are available in individual tea bags and 1# containers.

COCOA AND HOT CHOCOLATE

Cocoa and hot chocolate are made of roasted, shelled, cracked, and ground cacao beans. The cacao tree (*Theobroma cacao*) is grown in Africa, Mexico, Asia, and the West Indies. Like the coffee bean, the cacao bean develops its flavor through the roasting process. The flavor of cocoa varies according to the type of bean used. The cocoa butter is removed from the beans by grinding at 32°C (90°F). The paste, called "chocolate liquor," is cooled to a solid, called "bitter chocolate," which contains 50 to 60 percent fat.

The fat content of cocoa varies: breakfast or high-fat cocoa contains more than 25 percent fat; medium-fat cocoa contains 10 to 25 percent fat; and low-fat cocoa contains 6 to 7 percent fat. Cocoa and chocolate products contain about 6 percent of the amount of caffeine found in the average cup of coffee.

Cocoa can be seasoned or spiced, but this must be indicated on the label. It is marketed as natural process and Dutch process. Dutch process cocoa and chocolate have a more noticeable flavor and are much darker.

Instant cocoa is a mixture of cocoa, sugar, and an emulsifier. Soya lecithin is a natural product from soybeans used for its emulsifying properties; it helps sugar and fat particles form a mixture with smooth consistency. The beverage can be prepared for use without cooking by adding hot milk. Some varieties have dry milk solids in the mixture, so they must be reconstituted with water.

Federal standards establish identity for baking chocolate, sweet chocolate, semisweet chocolate, and milk chocolate.

FRUIT JUICES AND DRINKS

Fruit juices are available in a variety of forms. They may be fluid, frozen, or freeze-dried; single strength or concentrated; refrigerated or pasteurized to delay spoilage; or packaged in glass, cans, foil-lined paperboard, or vaporproof envelopes. Grade standards have been established for most fruit juices and these are similar to the standards used for fruits.

Fruit beverages and fruit bases are made in part from fruit juices and pulp. Juice drinks often contain little fruit or juice but derive most of their flavor from added acids, natural or synthetic flavoring materials, and other additives. Some drinks have no real fruit ingredients. The amount and quality of the fruit will be reflected in the price. Fruitades and imitation fruit juices are often enriched with vitamin C but do not contain the other nutrients found in real juices.

CARBONATED BEVERAGES

Sodas, or sweetened carbonated beverages, are available in a variety of flavors. Blended carbonated beverages include the colas and root beer. Fruit-flavored carbonated beverages include strawberry, orange, lemon, lime, and other single or combined fruit flavors. Both of these products have a low- or noncaloric counterpart made with artificial sweetener instead of sugar or high-fructose corn syrup. Some are caffeine-free as well. These drinks are sold in individual portion size cans and bottles. For quantity service they are available in 1-gallon and 5-gallon containers for use in a premix machine or in a syrup tank that is used in conjunction with tap water and carbon dioxide to blend carbonated beverages in a postmix machine.

WATER

Water is a colorless compound of hydrogen and oxygen and has no calories. Hard water contains high mineral concentrations like calcium and magnesium. Soft water contains higher sodium levels. The Environmental Protection Agency (EPA) regulates content of tap water. Most is disinfected with chlorine. In some areas, fluoride is added to prevent tooth decay. Minimum standards for safety and quality are set by the Food and Drug Administration (FDA) and match standards for municipal water supplies.

Sales of bottled water have risen despite the cost. It tastes better but may not be any safer or healthier than tap water. There are many types of bottled water and the composition varies.

Club Soda. Tap water that is filtered and artificially carbonated with carbon dioxide. Contains added salt and minerals.

Mineral Water. From either surface or groundwater. Any water containing minerals (not distilled). The International Bottled Water Association requires mineral water to contain not less than 500 parts per million of total dissolved solids. The more solids or minerals, the stronger the taste.

Natural Water. Not derived from a municipal system and has not been modified by adding or deleting minerals.

Purified (distilled) water. Completely demineralized by evaporation and recondensation processes. Flat tasting. Often used for medicinal purposes.

Seltzer. Same as club soda but has no added salts. May have juice and/or sugar added for flavor (adds calories).

Sparkling water. Contains carbon dioxide either naturally or added during bottling.

Spring water. Naturally flows out of the earth from a natural spring and is bottled at or near its source. Not altered by addition or deletion of minerals.

Still water. No bubbles and may or may not be processed. Comes from any source including municipal water supplies.

Well water. Extracted from a man-made hole in the ground that taps the water of an aquifer.

18 RECEIVING AND STORAGE

THE MAJOR OBJECTIVES of receiving and storage are to obtain the quality and quantity of merchandise ordered at the quoted price, to maintain adequate stocks of merchandise on hand, and to avoid loss through theft and spoilage. The attainment of these objectives can best be achieved by following certain proven practical principles, methods, and procedures.

The person responsible should know all aspects of the merchandise received. It should be checked for quality, quantity, specification, and price. All merchandise should be accompanied by an invoice, and that information should then be summarized on the purchase record. Constant follow-up and evaluation checks are necessary to ensure proper performance of the receiving procedure.

Management should impress upon all employees, especially those responsible for storage of food, the fact that food merchandise is money. There are many ways food quality can deteriorate: delays between receipt and storage, lack of adequate ventilation, long storage times, and failure to segregate foods correctly. Spot checks of the food, storage methods, storage temperature, and sanitation of storage areas must be done regularly by a supervisory person.

Meat is a very perishable food. Its palatability and high nutritive value must be protected by refrigeration from the time of slaughter until it is eaten. All fresh, smoked, and cooked meats should be refrigerated. Fresh meats should be stored at 1° to 3°C (33° to 36°F) and a relative humidity of 80 to 90 percent. Smoked and cooked meats may be stored at slightly higher temperatures, but not above 5°C (40°F). As soon as meat is delivered, its outside wrapping should be loosened or replaced with waxed paper to prevent drying. Ground meat and meat cut in small pieces spoil most rapidly. To prevent deterioration, meat should be stored tightly covered. When it is necessary to hold ground meat for more than 24 hours, it should be frozen.

Seafood is perishable, so it must be kept iced or refrigerated from the time it is caught until time for use. For longer storage, seafood items should be kept solidly frozen. Store fresh fish tightly wrapped in moistureproof, vaporproof material or in airtight containers in the refrigerator. Use within two days. Store frozen fish immediately to prevent thawing and deterioration of quality. Hold at $-25°C$ ($-10°F$) for not longer than six months. Prolonged storage adversely affects flavor, texture, and color of the product. Cooked fish that is stored in a covered container may be kept in the refrigerator up to three days. Live shellfish is shipped in ice and should be cooked immediately. The cooked meat should be refrigerated and used as soon as possible. Shuck oysters, clams, and scallops before freezing, allowing ½-inch airspace in the container. Crabs and lobsters should be cooked and chilled and have the meat removed from the shells before packaging and freezing. Use the frozen cooked meat within one month. Uncooked shrimp may be frozen in the shell or shelled. Use within three months.

Poultry, both cooked and raw, is very perishable. Unless properly cared for and refrigerated quickly, poultry will spoil readily. Fresh poultry should be stored at 0° to 5°C (32° to 40°F) for not longer than 24 hours. Frozen poultry may be kept in the wrapper in which it is received. It may be stored for not more than three days at 0° to 5°C (32° to 40°F) or six to nine months at $-20°C$ (0°F) or below. Refrigerate all cooked poultry promptly and use within 24 hours. Cover it well to prevent drying and absorption of odors from other foods.

Eggs are perishable and lose their quality rapidly if not properly cared for. They should be refrigerated immediately at 5° to 10°C (40° to 50°F). Eggs evaporate and absorb odors through the shell, so they should be kept covered and should not be stored near strong-flavored foods. Frozen eggs should be stored in their original container to prevent contamination. They should be held at $-20°C$ (0°F) or lower and should not be defrosted until ready for use. Dried eggs should be stored in airtight containers and refrigerated at 5° to 10°C (40° to 50°F).

Dairy products must be kept cold, covered, and clean. Milk and cream must be refrigerated immediately and held at 2° to 5°C (35° to 40°F) until just before use. Ice cream and frozen desserts must be held at −15° to −20°C (0° to 10°F).

Refrigerate natural cheese in the original wrapper. Cover cut surfaces tightly with foil or clear plastic wrap. Properly stored, natural cheeses will keep for several weeks. Store strong-flavored cheeses refrigerated in tightly covered containers. Refrigerate unripened cheeses immediately, as they are highly perishable; use soon after purchase. Freezing cheese is not generally recommended because the texture may alter after defrosting. If done, package tightly in moistureproof, vaporproof freezer wrapping and freeze at −20°C (0°F) or below. Thaw in the refrigerator and use as soon as possible. Frozen cheese will keep several months.

Fruit can be held in storage considerably longer by proper control of temperature, moisture, and respiration conditions. Fruit should be carefully watched and any showing signs of spoilage should be removed. Fruit should not be refrigerated until ripened. Low temperatures retard enzyme action and help to hold fruit at the desired state of ripeness. When ripened bananas are held in low-temperature storage, the skin turns brown. Berries are among the most perishable fruits. They crush easily and should have a minimum of handling. For this reason they should be stored in the small boxes in which they are received. These containers are well ventilated and thereby help retard the growth of mold.

Vegetables should be eaten as soon as possible after they are delivered. They should be examined carefully to see that they are not bruised or substandard. They should not be washed until time for preparation. Vegetables held for later use should be kept between 5° and 10°C (38° and 50°F) to prevent deterioration. Root vegetables and tubers being held for long periods must have both temperature and humidity controlled. When potatoes are stored at temperatures below 10°C (50°F), much of the starch is changed to sugar, making them unsatisfactory for use as potato chips and French-fried potatoes.

The cooking quality is best when potatoes are stored between 10° and 15°C (50° and 60°F). For long storage times, temperatures of 5° to 10°C (38° to 50°F) are recommended to prevent sprouting. The change from starch to sugar caused by storage at such low temperatures can be reversed by storing for a period of time above 10°C (50°F). The flavor of parsnips is improved by storage at a low temperature, which stimulates the change of the starch to sugar.

Staples requiring room temperature storage, such as canned fruits and vegetables, soup, juice, and cereal, should be clean and dry and stored in a well-ventilated area. Optimum temperature is 10° to 20°C (50° to 70°F). Temperatures above this reduce the shelf life of the food being stored. Most states require that food be stored above the floor on shelves or pallets to aid in cleaning, prevent water damage, and improve circulation of air.

Spices should be kept in tightly closed containers since much of their aroma and flavor are lost on exposure to air. Herbs and ground spices have a shorter shelf life than whole spices. Deterioration is more rapid at higher temperatures. Flavorings should be kept tightly sealed to prevent evaporation.

Foodservice costs are influenced by many persons and the performance of many activities from the time merchandise is purchased until it is eaten. The responsibilities of key personnel involved in these activities need to be clearly defined. The procedures for effectively carrying out and evaluating the performance of these duties must be developed. If the basic operating procedures are efficient, the proper foundation will be laid for effectively controlling costs at later stages of food processing.

The primary objective of food purchasing is to obtain the best quality of merchandise, based on established specifications, at the best possible price and to receive and store these products in an efficient manner. Knowledge of accounting procedures and product specifications will allow the purchaser to make the decisions necessary for a successful foodservice operation.

INDEX

Acesulfame K, 132
Allspice, 120
Amaranth, 116
Anise, 123
Apple pie spice, 124
Apples, 76–77
Apricots, 77
Artichokes, 88
Asparagus, 88
Aspertame, 133
Avocados, 77

Bacon, 36
Bananas, 78
Barbecue spice, 125
Barley, 108–109
Basil, 122
Bass, 45
Bay leaves, 122
Beans, 88–89
Beef, 23–33
Beets, 89
Beverages, 135–142
Blackberries, 78
Blueberries, 78
Blue cheese, 70–71
Bluefish, 46
Bok choy, 89
Brick cheese, 70–71
Brie cheese, 71–72
Broccoflower, 89
Broccoli, 89–90
Brown sugar, 129
Brussels sprouts, 90
Buckwheat, 116–117

Budget, 10, 13
Bulgar, 113
Butter, 67–68
Buttermilk, 64

Cabbage 90
Camembert cheese, 71–72
Can cutting, 6, 9
Cane syrup, 130
Can sizes, 104, 107
Cantaloupe, 78–79
Capon, 51
Capsicum pepper, 120
Caraway seeds, 124
Carbonated beverages, 140–141
Cardamom seeds, 124
Carrots, 91
Catfish, 46
Cauliflower, 91
Celery, 91–92
Celery flakes, 126
Celery seeds, 124
Cereal products, 108–119, 146
Cheddar cheese, 69–70
Cheese, 69–74
Cheese, imitation, 74
Cheese, low-fat, 74
Cheese, low-sodium, 73–74
Cheese food, 73
Cheese spread, 73
Cherries, 79
Chervil, 122
Chicken, 51–53
Chili powder, 125
Chives, 97, 122, 127

Cinnamon, 121
Clams, 49
Cloves, 121
Cocoa, 139–140
Cod, 45
Coffee, 135–138
Colby cheese, 70
Cold pack cheese, 73
Collards, 92
Condensed milk, 65
Convenience food, 6
Coriander, 122, 124
Corn, 92, 109–110
Corn cereals, 110
Corn flour, 109
Cornmeal, 109
Corn oil, 110
Cornstarch, 110
Corn sugar, 129
Corn syrup, 130
Cost comparisons, 5
Cottage cheese, 71–72, 74
Couscous, 116
Crab, 49
Crab boil, 125, 126
Cranberries, 79
Cream, 66, 145
Cream cheese, 71–72
Cucumber, 92–93
Cumin, 124
Curry powder, 125–126
Cyclamate, 133–134

Dairy products, 63–68, 145
Dates, 92
Delivery frequency, 6
Dill, 123, 124
Direct purchases, 8
Dried eggs, 61–62
Duck, 51, 55

Edam cheese, 70–71
Egg grades, 57–58
Eggplant, 93
Egg products, 60–62

Eggs, 57–62, 144–145
Eggs, dried, 61–62
Eggs, frozen, 60–61
Egg sizes, 58–59
Endive, 93
Escarole, 93–94
Evaporated milk, 65

Farina, 115
Fennel, 124
Fenugreek, 124
Feta cheese, 71–72
Filled milk, 65
Flatfish, 45
Flavored salt, 126
Flavorings, 127, 146
Flounder, 45
Fluke, 45
Food, Drug, and Cosmetic Act, 4
Food and Drug Administration, 43, 51
Food colorings, 127
Food cost, 4
Food inventory, 11, 16
Food safety, 5
Formal buying, 6
Frozen meat, 30
Fructose, 131
Fruit, canned, frozen, and dried, 104–107, 145–146
Fruit, fresh, 75–86, 145
Fruit bases, 140
Fruit beverages, 140
Fruit grades, 75–86
Fruit juices, 140, 145

Garlic, 121, 127
Ginger, 121
Goose, 51, 56
Gooseberries, 92
Gouda cheese, 70–71
Grains, 108–117
Grapefruit, 79–80
Grapes, 80
Greens, 94
Grits, 109

Ground beef, 29
Grouper, 45
Guinea, 51, 56

Haddock, 45
Hake, 45
Halibut, 45
Ham, 35–36
Herbs, 122–123, 146
Herb seasoning, 126
Hominy, 109–110
Honey, 131–132
Horseradish, 123

Ice cream, 66–67
Ice milk, 67
Imitation cheese, 74
Individually quick frozen (IQF) meat, 30
Informal buying, 7
Institutional Meat Purchase Specification (IMPS), 27
Irradiation, 5, 23, 34, 43, 51, 75, 106, 108, 120
Italian seasoning, 126

Jam, 132–133
Jelly, 132–133
Juices, 140, 146

Kale, 94
Kiwi fruit, 80

Lamb, 40–42
Legumes, 108, 117–118
Lemons, 81
Lettuce, 94–95
Limburger cheese, 71–72
Limes, 81
Lobster, 48–49
Lupini pasta, 116

Mace, 121
Mackerel, 46
Mandarins, 85
Maple sugar, 130
Maple syrup, 131
Margarine, 68
Marjoram, 123
Marmalade, 132
Meal census, 12, 20
Meal cost, 12
Meat, 23–33, 34–39, 40–42, 143–144
"Meat Buyer's Guide," 28
Meat grades, 24, 40
Meat Inspection Act, 23
Meat specifications, 27
Melons, 81–82
Menu, 3, 10, 14, 15
Milk, 63–65, 145
Millet, 110
Mint, 122
Mixed pickling spices, 126
Mixed vegetable flakes, 127
Molasses, 130
Monkfish, 47
Monterey Jack cheese, 70–71
Monthly summary, 12, 22
Mozzarella cheese, 70–71
Muenster cheese, 70–71
Mushrooms, 95–96
Mustard, 125
Mustard greens, 96

Nectarines, 82
Neufchatel cheese, 71–72
Nonfat dry milk, 65
Nonnutritive sweeteners, 133–134

Oatmeal, 110
Oats, 110
Okra, 96
Onions, 96–97, 127
Open buying, 7
Oranges, 82–83
Ordering, 3
Oregano, 123

Oysters, 50

Paprika, 121
Parmesan cheese, 69
Parsley, 97
Parsley flakes, 126
Parsnips, 98
Par stock, 7
Partridge, 56
Pasta, 115–116
Peaches, 83
Pears, 83
Peas, 98
Pepper, 121
Pepper flakes, 127
Peppers, 98–99
Pheasant, 56
Pickling spices, mixed, 126
Pineapple, 83–84
Plaice, 45
Plums, 84
Pollock, 45
Polyalcohols, 133
Poppy seeds, 125
Pork, 34–39
Portion cost, 12, 21
Potatoes, 99
Poultry, 51–56, 144
Poultry seasoning, 126
Poussin, 56
Preportioned meat, 27
Preserves, 132–133
Price quotations, 11, 17
Process cheese, 72–73
Provolone cheese, 70
Prunes, 84
Pumpkin, 100
Purchase order, 11
Purchase record, 11, 18, 19

Quail, 56
Quality grade, meat, 23
Quinoa, 117

Radishes, 100

Raspberries, 84
Receiving and storage, 143–146
Recipes, 3
Red snapper, 46
Restructured meat, 31
Rice, 111–112
Ricotta cheese, 71–72
Rock cornish game hens, 51
Romano cheese, 69
Roquefort cheese, 70–71
Rosemary, 123
Rutabagas, 100
Rye, 112
Rye flour, 112

Saccharin, 133
Saffron, 121–122
Sage, 123
Salmon, 46
Sausages, 36–37
Savory, 123
Scallops, 48
Scrambled eggs, 61
Seafood, 43–50, 144
Seafood seasoning, 126
Seasoned salt, 126
Seasonings, dry vegetable, 126–127
Seeds, 124–125
Semolina, 115
Sesame seeds, 125
Shallots, 97
Sherbet, 67
Shrimp, 47
Shrimp spice, 126
Soba noodles, 117
Sole, 45
Sorghum syrup, 130
Sour cream, 65
Soybeans, 117–118
Soybean sprouts, 18
Soy flour, 117
Soy grits, 118
Soy meal, 30, 31
Soy milk, 118
Soy oil, 117–118
Spices, 120–122, 146

Squab, 56
Squash, 101
Standing orders, 7
Storage, 6
Strawberries, 84–85
Sugar, 128–130
Surimi, 50
Sweetening agents, 128–134
Sweet potatoes, 101
Swiss cheese, 69–70
Swordfish, 46
Syrups, 130–133

Tangelos, 85
Tangerines, 85
Tarragon, 123
Tea, 138–139
Temple oranges, 85
Textured vegetable protein (TVP), 30, 118
Thyme, 123
Tomatoes, 85, 101–102
Triticale, 112
Trout, 46
Tuna, 44, 46
Turbot, 45
Turkey, 51, 53–55
Turmeric, 122
Turnips, 102

USDA, 23, 24, 26

Variety meats, 30, 35
Veal, 27, 33
Vegetable oil, 119
Vegetable protein product (VPP), 30, 118
Vegetables, canned, frozen, and dried, 104–107, 146
Vegetables, fresh, 87–103, 145–146

Water, 141–142
Watercress, 102
Water ice, 67
Watermelon, 85
Wheat, 112–116
Wheat cereal, 113
Wheat flour, 114–115
Wheat germ, 113
Wheat sprouts, 115
Whipped topping, 66
Wurld wheat, 113

Yield grades, 26, 34, 41
Yogurt, 64–65